Laboratory Manual/Workbook

GENERAL SCIENCE

Patricia A. Watkins
Science Curriculum Specialist
and former Science Teacher
Northside Independent School District
Instructor, Department of Education
Trinity University
San Antonio, Texas

Cesare Emiliani
Chairperson, Department of Geology
Professor of Geological Sciences
University of Miami
Coral Gables, Florida

Christopher J. Chiaverina
Physics Teacher
Barrington High School
Barrington, Illinois

Christopher T. Harper
Instructor in Science
Phillips Exeter Academy
Exeter, New Hampshire

David E. LaHart
Senior Instructor
Florida Solar Energy Center
Cape Canaveral, Florida

Harcourt Brace Jovanovich, Publishers
Orlando   San Diego   Chicago   Dallas

Writers and Reviewers

Lowell J. Bethel, Ed.D.
Associate Professor of Science Education
Science Education Center
University of Texas at Austin
Austin, Texas

Richard F. Davis
Reading Specialist
Lee County Schools
Lee County, Florida

Patti Day-Miller
Reading Consultant
Angola, Indiana

Ralph DeFrehn
Science Teacher
North Fort Myers High School
North Fort Myers, Florida

Sylvia D. Geshell
Science Consultant
Missoula, Montana

Dianne I. Hillman
Earth Science Teacher
Cherokee High School
Lenape Regional High School District
Marlton, New Jersey

Charles N. Kish
General Science Teacher
Saratoga Springs Junior High School
Saratoga Springs, New York

Karen M. Nein
Science Consultant
Englewood, Colorado

Walter P. Seibel
Supervisor
Mathematics and Science Department
Cherokee High School
Lenape Regional High School District
Marlton, New Jersey

Rajee Thyagarajan
Physics Teacher
Health Careers High School
San Antonio, Texas

Copyright © 1989 by Harcourt Brace Jovanovich, Inc.

All rights reserved. No part of this publication may be reproduced or transmitted in any form or by any means, electronic or mechanical, including photocopy, recording, or any information storage and retrieval system, without permission in writing from the publisher.

Requests for permission to make copies of any part of the work should be mailed to: Permissions, Harcourt Brace Jovanovich, Publishers, Orlando, Florida 32887

Acknowledgment: From *Biology,* Laboratory Investigations, Annotated Teacher's Edition. Copyright © 1986 by Harcourt Brace Jovanovich, Inc. Reprinted by permission of Harcourt Brace Jovanovich, Inc.

Printed in the United States of America

ISBN 0-15-364308-0

Contents

How to Use the Book 5
How to Conduct an In-class Investigation 7
How to Write a Laboratory Report 10
Apparatus 12
How to Use a Bunsen Burner, Graduate, and Balance 13
Laboratory Safety for the Student 15

Investigations 17

1. Density 19
2. Friction 23
3. Conservation of Matter 25
4. Filtering a Mixture 27
5. Designing Electron Energy Shells 31
6. Baking Soda Reaction 33
7. Domino Reaction 35
8. Polymer Properties 37
9. Calculating the Speed of Sound 41
10. Defining Work 43
11. Phase Change of Water 47
12. Electromagnetism 51
13. Observing Fields of View 55
14. Remote Control 59
15. The Changing of the Seasons 61
16. How Does a Planet Remain in Its Orbit? 65
17. Constellations 67
18. Making a Fossil 71
19. Pangaea, the Supercontinent 73
20. Condensation of Water 77
21. Weathering 79
22. What Causes Ocean Currents? 83
23. A Classification System for Leaves 87
24. Random Assortment of Genetic Traits 91
25. Variation Within a Species 95
26. Predator and Prey Relationships 99
27. Lung Capacity 103
28. Ecological Treasure Hunt 107
29. Corridor Problem Solving 109

Vocabulary & Concepts

Vocabulary

Chapter	1	115
Chapter	2	117
Chapter	3	119
Chapter	4	121
Chapter	5	123
Chapter	6	125
Chapter	7	127
Chapter	8	129
Chapter	9	131
Chapter	10	133
Chapter	11	135
Chapter	12	137
Chapter	13	139
Chapter	14	141
Chapter	15	143
Chapter	16	145
Chapter	17	147
Chapter	18	149
Chapter	19	151
Chapter	20	153
Chapter	21	155
Chapter	22	157
Chapter	23	159
Chapter	24	161
Chapter	25	163
Chapter	26	165
Chapter	27	167
Chapter	28	169
Chapter	29	171

Concepts

Chapter	1	116
Chapter	2	118
Chapter	3	120
Chapter	4	122
Chapter	5	124
Chapter	6	126
Chapter	7	128
Chapter	8	130
Chapter	9	132
Chapter	10	134
Chapter	11	136
Chapter	12	138
Chapter	13	140
Chapter	14	142
Chapter	15	144
Chapter	16	146
Chapter	17	148
Chapter	18	150
Chapter	19	152
Chapter	20	154
Chapter	21	156
Chapter	22	158
Chapter	23	160
Chapter	24	162
Chapter	25	164
Chapter	26	166
Chapter	27	168
Chapter	28	170
Chapter	29	172

How to Use the Book

To the Student

The study of science can be exciting, as well as informative. To increase your understanding of concepts presented in **GENERAL SCIENCE**, this *Laboratory Manual/Workbook* provides Investigations and worksheets that will enhance your learning experiences.

Investigations are closely correlated with chapter concepts, and their use will make your studies more meaningful and worthwhile. These Investigations will help you:

- *discover* facts relating to scientific concepts.
- *use* techniques associated with the tools of science.
- *correlate* observed events with known scientific theories.
- *formulate* hypotheses about scientific events.
- *use* a scientific method to solve problems in the laboratory.
- *investigate* concepts presented in the textbook.

Each Investigation has five parts.

- The **Purpose** is a statement of what will be done in the Investigation.
- The **Materials** section is a list of all apparatus, equipment, chemicals, or specimens needed to complete the Investigation. Materials are listed in the order in which they are used in the Investigation. Where appropriate, and at the top of the list, safety symbols are included. Safety symbols alert you to materials or procedures that require special care. These symbols are the same as those found in your textbook.

The symbols are:

safety goggles needed for eye protection

laboratory apron needed for protection of clothing

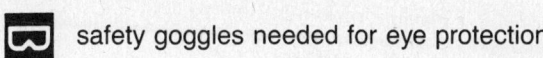 sharp/pointed object; use care not to cut yourself

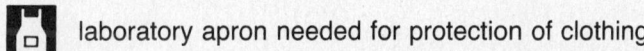 heat source, open flame; use care not to burn yourself or to ignite clothing, papers or other materials

chemical hazard; may be poisonous or corrosive

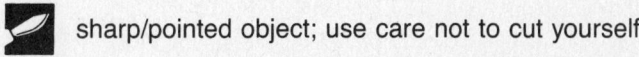 electrical hazard; use care to avoid electrical shock

- The **Procedure** includes step-by-step instructions for carrying out the Investigations and questions relating to these steps. Data is collected in this section. Caution statements are included for any procedure that is considered potentially hazardous. Pay special attention to these statements and to the safety symbols in each Investigation.

- The **Results** section provides you with space to record the data collected in the **Procedure**.

- The **Conclusions** section is a list of questions that require you to use the knowledge gained in the Investigation. Sometimes you may need to use your textbook as a resource to help you answer these questions.

An **Applications** section is included where appropriate. **Applications** are intended to show the practical nature of the Investigations and to provide you with an opportunity to apply the concepts to the world around you.

Experiences in the laboratory can be enjoyable and worthwhile if you follow instructions and take proper precautions.

How to Conduct an In-class Investigation

In order to be successful in the science laboratory, you must follow some simple rules. These rules apply equally to laboratory investigations and to field work. The first thing you, as a student, must do before beginning an investigation is to outline what you intend to do. In many cases, you will be following specific instructions from your teacher or from the *Laboratory Manual/Workbook*. Be sure to read over any written instructions carefully before you begin any investigation. Make sure you understand all instructions. If you have any questions, ask your teacher for help before you begin. If you are given oral instructions, carefully write down all information as accurately as possible. If you are designing and conducting an independent investigation, such as a science fair project, you must plan very carefully before beginning.

Once you understand the procedure you will be using, the next thing you must do is list all the equipment and supplies you will need to complete the investigation. Check the procedure carefully to be sure all equipment is included on your list. Before you begin your investigation, gather all the materials you will need. Be sure that you know how to use the equipment that is necessary for your investigation. If you are unsure of any operating procedure, ask your teacher for help.

While following the procedure outlined for an investigation, you must make certain that all results are recorded exactly and precisely. One of the most important aspects of scientific investigation is accuracy. This means that if you are recording masses from a balance, you must record the *exact* mass indicated on the balance. If you are counting how many cells are present in the field of vision of your microscope, you must actually count them and not guess how many are there.

Your data is all the information that you have gathered from an investigation. The type of data collected depends upon the activity. It may be a series of numbers, or a set of color changes, or a list of scientific names. No matter which type of data is collected, all of it must be treated carefully to ensure accurate results. Sometimes the data may seem to be incorrect, but even then, it is important and should be recorded accurately. Data that seems to be "wrong" is, most probably, the result of an experimental error.

There are many ways in which to record data. These methods include data tables, graphs, and diagrams.

Data Tables

Data tables are probably the most common means of recording data. Although prepared data tables are provided in the *Laboratory Manual/Workbook*, it is important that you also be able to construct your own. The best way to do this is to choose a title for your data table and then make a list of the types of data to be collected. This list will become the headings for your data columns. When filling out the data table, always be careful to place the information in the correct column or row. This data is the basis for all your later interpretations and analyses. You can always ask new questions about the data, but you cannot get new data without repeating the experiment.

Graphing

After data is collected, you must determine how to display it. One way of showing your results is to use a graph. Two types of graphs are commonly used, the line graph and the bar graph. In a line graph the data are arranged so that two variables are represented as a single point. For example, if you collected data on plant growth that included both the time of growth and the amount of growth, you could record your data in a table like this.

Plant Growth Data

Time in Days	Height of Plant (cm)
1	10
2	11
3	11.5
4	12
5	12.5

You could easily make a graph of this data. The first step is to draw and label the axes. Before you do this, however, you must decide which column of data should be on the x, or horizontal, axis and which should be on the y, or vertical, axis. The x axis commonly represents the independent variable. That is, the data that could be present even if the rest of the data were not. For this experiment, the independent variable is "Time in Days." Time exists regardless of whether plants are present or not. Therefore, the y axis would be the height of the plants, the dependent variable. Next, you must choose the scale for the axes. You want the graph to take up as much of the paper as possible, since large graphs are much easier to read than small ones. You must choose a scale for each axis that utilizes the largest amount of graph paper. Remember, once you choose the interval for the scale, for example, the number of days each block represents on the x axis, you cannot change it. You cannot say that block one represents one day and block two represents 10 days. If you change the scale, your graph will not accurately represent your data.

The next step is to mark the points for each pair of numbers. When all points are marked, draw the best straight or curved line between them. Remember that you do not "connect the dots" when you draw a graph, you must draw the best line possible. The best line may not include every point in the data.

You may choose to represent your data by using a bar graph. The first steps are similar to those for the line graph. You must choose your axes and label them first. The independent variable remains on the x axis and the dependent variable is on the y axis. However, instead of plotting the points on the graph, you represent the dependent variable as a bar from the x axis to the point of intersection with the y axis coordinate. Using the data on plant growth, for example, on day one the height of the plant in the sample data is 10 cm. On your graph you would make a bar (usually one block in width) to the height of the 10 cm mark on the y axis. Each bar is a separate piece of data, unlike the points in the line graph.

Diagrams

In some cases, the data you must represent is not numerical. That means that it cannot be put into a data table or be graphed. The best way to represent this data is to draw and label it. To do this you simply draw what you see and label as many parts or structures as possible. This technique is especially useful in investigations that involve observation of living or preserved specimens. Remember, you do *not* have to be an artist to make good laboratory diagrams.

There are several things you need to remember as you make your laboratory diagrams. First, make the diagram large enough so that it may be easily studied. Include all the visible structures in your diagram. Second, your diagrams should also show the spacing between the parts of the specimen in proportion to its actual appearance. Size relationships are important in understanding and interpreting observations. Third, in order for your drawings to be the most useful to you, you need to label them. All labels should be clearly and neatly printed. Lines drawn from labels to the corresponding parts should be straight. Use a ruler. Label lines should *never* cross each other. Fourth, be sure to title all drawings. Someone who looks at your drawing should be able to identify the specimen. Remember, neatness and accuracy are the most important parts of any laboratory drawing.

How to Write a Laboratory Report

There are two different types of laboratory reports that you may be asked to write. The first is a report of a laboratory investigation in which the results and your interpretation of the results are the most important items required by your teacher. This type of investigation is usually found in a laboratory manual, where the procedure is already outlined for you. Such reports would contain the following parts.

Title This is the name of the laboratory investigation you are doing. In an investigation from a laboratory manual, the title will be the same as the title of the investigation.

Hypothesis The hypothesis is what you think will happen during the investigation. It is often posed as an "If . . . then" statement. For example: If sulfuric acid is added to sugar, then the sugar will be broken down into its chemical components.

Materials This is a list of all the equipment and other supplies you will need to complete the investigation. In investigations taken from a laboratory manual, the materials are generally listed for you.

Procedure The procedure is a step-by-step explanation of exactly what you did in the investigation. Investigations from laboratory manuals will have the procedure carefully written out for you, all you need to do is to read it very carefully. Often, in laboratory manuals, there will be questions in the procedure section that will help you understand what is happening in the investigation.

Results Your data is what you have observed. It is often recorded in the form of tables, graphs, and diagrams.

Conclusions This is the most important and difficult part of the investigation. It explains what you have learned. You should include everything you have learned; you should explain any errors you made in the investigation; and you should evaluate your hypothesis. Keep in mind that not all hypotheses will be correct. This is normal. You just need to explain why things did not work out the way that you thought they would. In laboratory manual investigations, there will be questions to guide you in analyzing your data. You should use these questions as a basis for your conclusions.

In some cases, you might be required to do an independent project. You may design your own investigation for a science fair project, or your teacher may have you design an investigation to perform in class. The report for this type of investigation should include two sections that are not included in the previous type of report. These sections are **Introduction** and **References**. In order for a laboratory report on an independently designed experiment to be complete, you must now include an introduction and a reference section. They should be included in your report in the following order.

Title

Introduction The introduction should include a clear, simple statement of your purpose. In addition, the introduction should include a discussion of the important ideas that led you to design and perform the experiment. For example, you could include such things as why you are doing this investigation, what is interesting about the topic to be investigated, and what information you have already gathered about the topic. In order to prepare a good introduction, you will need to do library research on the topic. Be sure to use proper footnoting methods when you use ideas from any reference source.

Hypothesis

Materials

Procedure

Results

Conclusions

References List all the reference materials used to begin and to complete the project. Be sure to use complete citations, including author, title, date of publication, and place of publication. Your teacher will give you the format preferred for the type of investigation you are doing.

Remember that a good laboratory report takes time. Do not wait until the night before the report is due to begin work on it.

Apparatus

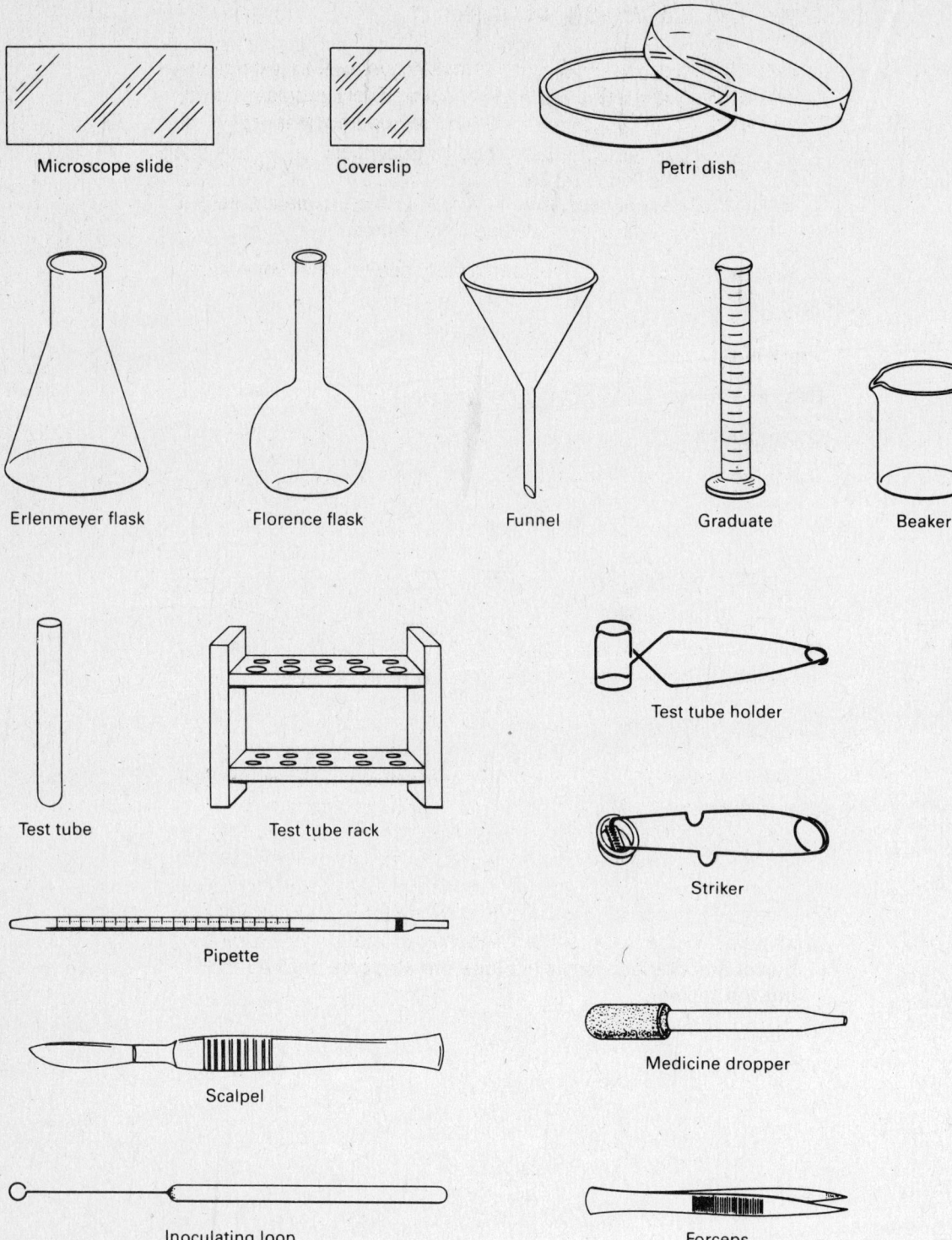

12 Apparatus

How to Use a Bunsen Burner, Graduate, and Balance

Lighting a Bunsen Burner

1. Before lighting the burner, observe the locations of fire extinguishers, fire blankets, and sand buckets. Wear safety goggles, gloves, and an apron. Tie back long hair and roll up long sleeves.

2. Turn the gas full on by using the valve at the laboratory gas outlet.

3. Make a spark with a striker. If you are using a match, hold it slightly to the side of the opening in the barrel.

4. Adjust the air ports until you can clearly see an inner cone within the flame.

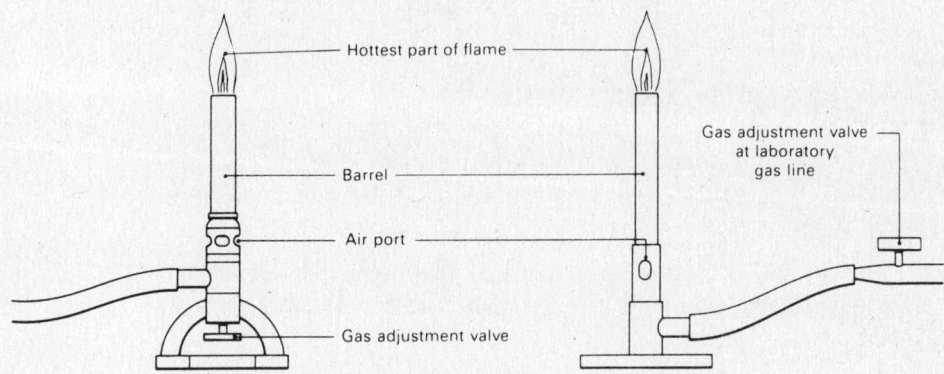

5. Adjust the gas flow for the desired flame height by using the gas adjustment valve either on the burner or at the gas outlet.
CAUTION: If the burner is not operating properly, the flame may burn inside the base of the barrel. Carbon monoxide, an odorless gas, is released from this type of flame. Should this situation occur, turn off the gas at the laboratory gas valve immediately. Do not touch the barrel of the burner. After the barrel has cooled, partially close the air ports before relighting the burner.

Measuring Volume in a Graduate

1. Set the graduate (graduated cylinder) on a level surface.
2. Carefully pour the liquid you wish to measure into the cylinder. Notice that the surface of the liquid has a lens-shaped curve, the *meniscus*.
3. With the surface of the liquid at eye level, read the measurement at the bottom of the meniscus.

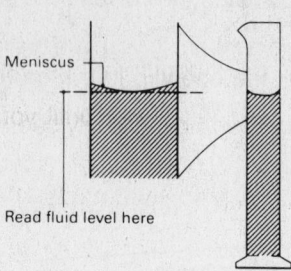

Using a Triple-beam Balance

1. Make sure the balance is on a level surface. Use the leveling screws at the bottom of the balance to make any necessary adjustments.
2. Place all the counterweights at zero. The pointer should be at zero. If it is not, adjust the balancing knob until the pointer rests at zero.
3. Place the object you wish to measure on the pan.
   CAUTION: Do not place hot objects or chemicals directly on the balance pan as they can damage its surface.
4. Move the largest counterweight along the beam to the right until it is at the last notch that does not tip the balance. Follow the same procedure with the next largest counterweight. Then, move the smallest counterweight until the pointer rests at zero.
5. Total the readings on all beams to determine the mass of the object.
6. When weighing crystals or powders, use a filter paper. First weigh the paper, then add the crystals and powders, and reweigh. The actual weight is the total weight less the weight of the paper. When weighing liquids, first weigh the empty container, then the liquid and container. The actual weight is the total weight less the weight of the container.

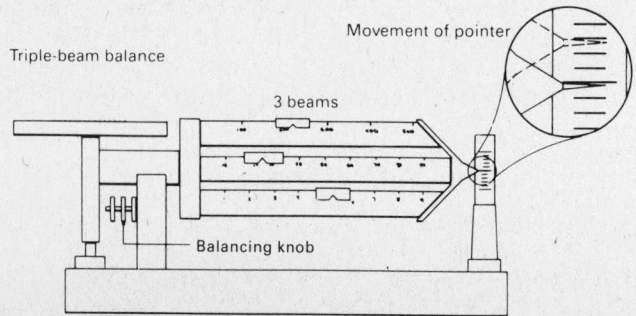

14   How to Use a Bunsen Burner, Graduate, and Balance

Laboratory Safety for the Student

Below is a list of precautions you should take *every* time you enter the laboratory. Using these precautions will make your laboratory experience safer and, therefore, much more enjoyable.

General Laboratory Procedures

1. Never "horse around" in the laboratory.
2. Never play with laboratory equipment or materials.
3. Always follow instructions and wait until you are told to begin before starting any investigation.
4. Never carry out unassigned experiments.
5. Never eat or taste anything in the laboratory. This includes food, drinks, and gum, as well as chemicals found in the laboratory.
6. Wash your hands after *every* experiment.
7. Keep all nonessential items away from the work area.
8. Keep your work area clean. Dispose of waste materials in appropriate containers.
9. Turn off any gas jets or any electrically operated equipment as soon as you are finished with them.
10. Report all injuries or accidents to your teacher immediately.
11. Never use broken or cracked glassware.
12. Always wear shoes in the laboratory. Sandals are not suggested.
13. Tie back long hair and restrict any loose clothing.
14. Wear safety goggles, laboratory aprons, and gloves when instructed to do so.

Working with Fire or Heat

1. Whenever possible, use a hot plate for heating. Use a gas burner only when specifically instructed to do so.
2. When heating materials, be sure the containers are made of heatproof glass.
3. Never point a heated container at anyone. Be especially careful with heated test tubes.
4. Turn off the heat source as soon as you are finished with it. Do not move it until it has cooled. Leave a note to indicate that a hot plate has been recently used, since it is difficult to tell without touching it.
5. Keep flammable materials away from open flames.
6. Use test tube holders, tongs, or heavy gloves to handle hot items.
7. Do not put your face or hands over any boiling liquid.

Working with Acids and Bases

1. Acids and bases are very caustic and are chemically active. Never pour water into an acid or base. Always pour acids and bases into water.

2. Should you get acid or base on your skin, dilute the chemical with running water and notify your teacher immediately.

Working with Electrical Equipment

1. Be careful with electrical cords. Never leave them where someone might trip over them. Keep the cords away from heat and water.

2. Never touch cords or electrical outlets with wet hands.

3. Grasp the plug when disconnecting an electrical cord from the outlet. Do not pull on the cord.

4. Turn off all equipment as soon as you are finished with it.

Working with Specimens—Living and Preserved

1. Treat all living things with respect. Do not grab or squeeze animals. Do not pick plants unless necessary.

2. Most small mammals have sharp teeth. Treat animals with care and respect.

3. When working with bacteria, be sure to clean your work area with disinfectant before and after an investigation.

4. Wash your hands after handling any specimen.

5. When dissecting, always place the specimen in a dissecting tray or on a prepared surface. Do not hold a specimen in your hand while dissecting it.

6. Sharp instruments such as scalpels, probes, and scissors are used in dissecting. Always use them carefully. They can cause severe cuts if they are mishandled.

Investigations

To complement the Investigations in the textbook, these Investigations are designed to extend hands-on laboratory experiences in science.

1	Density	19
2	Friction	23
3	Conservation of Matter	25
4	Filtering a Mixture	27
5	Designing Electron Energy Shells	31
6	Baking Soda Reaction	33
7	Domino Reaction	35
8	Polymer Properties	37
9	Calculating the Speed of Sound	41
10	Defining Work	43
11	Phase Change of Water	47
12	Electromagnetism	51
13	Observing Fields of View	55
14	Remote Control	59
15	The Changing of the Seasons	61
16	How Does a Planet Remain in Its Orbit?	65
17	Constellations	67
18	Making a Fossil	71
19	Pangaea, the Supercontinent	73
20	Condensation of Water	77
21	Weathering	79
22	What Causes Ocean Currents?	83
23	A Classification System for Leaves	87
24	Random Assortment of Genetic Traits	91
25	Variation Within a Species	95
26	Predator and Prey Relationships	99
27	Lung Capacity	103
28	Ecological Treasure Hunt	107
29	Corridor Problem Solving	109

Name _____ Class _____ Date _____

CHAPTER
1
Investigation

Science and Discovery

Density
Will equal volumes of different substances have the same masses and densities?

PURPOSE

To find the masses of equal volumes of two different substances
To determine the densities of the two substances

MATERIALS

50-mL beakers (2)   graduate
balance   farina
white sand

PROCEDURE

1. Place an empty, clean beaker on one pan of the balance, and find its mass by adding weights to the other pan. Record the mass in the Data Table.

2. Measure 30 mL of sand in the graduate, and pour it into the beaker. Find the total mass of the beaker and the sand. Record the mass in the Data Table.

3. Place another empty beaker on the balance, and find its mass. Record the mass in the Data Table.

4. Measure 30 mL of farina in the graduate, and pour it into the beaker. Find the total mass of beaker and the cereal. Record the mass in the Data Table.

5. By subtracting the mass of the empty beaker from the mass of the beaker and the sand, find the mass of the sand alone. Record this mass in the Data Table. Repeat this procedure to obtain the mass of the farina.

6. Determine the mass of 1 mL of each substance by dividing the mass of each by the volume (30 mL). The mass of 1 mL of a substance is that substance's density. Enter the density of the sand and of the cereal in the Data Table.

(continues)

Name _____ Class _____ Date _____

Chapter 1   Investigation   (continued)

RESULTS

Data Table

Substance	Mass of empty beaker (g)	Mass of beaker and substance (g)	Mass of substance (g)	Volume of substance (mL)	Density of substance (g/mL)
Sand					
Farina					

CONCLUSIONS

1. Which had the greater mass, the 30 mL of sand or the 30 mL of farina?

   _____

   _____

   _____

2. Which had the greater density, the 30 mL of sand or the 30 mL of farina?

   _____

   _____

   _____

3. How is density affected by mass and volume?

   _____

   _____

   _____

4. How might the same procedure be used to determine whether water or oil has a greater density?

   _____

   _____

   _____

(continues)

Name _____ Class _____ Date _____

Chapter 1   Investigation   (continued)

APPLICATIONS

What other method can be used to demonstrate differences in densities of liquids?

_____
_____
_____
_____

Name _____ Class _____ Date _____

CHAPTER **2** | **Science and Modern Technology**
Investigation

Friction

Does the nature of a surface affect the force of friction?

PURPOSE

To compare the forces needed to slide an object on a rough surface and on a smooth surface

MATERIALS

textbook with string attached (or 8 cm × 8 cm block of wood 12 to 16 cm long with string attached)
spring scale
wooden board 24 cm × 96 cm (one side made rough with an overlay of very rough sandpaper and the other side covered with cellophane)

PROCEDURE

1. Attach the book (or block of wood) to the spring scale. Place the book on the side of the board covered with cellophane. Holding the handle on the spring scale, pull with a steady, slow speed. When the book is sliding at a uniform speed, observe the reading on the spring scale. Record the reading in the Data Table.

2. Repeat this procedure two more times to get data for three trials. Record the reading from the spring scale each time.

3. Turn the board over so the sandpaper surface faces up. Repeat step 1 three times on the rough side, and record the reading from the spring scale each time.

(continues)

Name _____ Class _____ Date _____

Chapter 2 Investigation (continued)

RESULTS

	Data Table Force (spring-scale reading)			
	Trial 1	Trial 2	Trial 3	Average
Smooth surface				
Rough surface				

1. Determine the average of the three readings for each surface. Record the averages in the data table. The spring scale measures the force required to move the object on the surface.

2. Compare the force needed for the rough surface with the force needed for the smooth surface.

CONCLUSIONS

1. What are some examples of smooth surfaces being necessary to the efficient use of energy?

   _____
   _____
   _____

2. What are some examples of rough surfaces being necessary to the efficient use of energy?

   _____
   _____
   _____

3. What would be the benefit of making road surfaces slightly rough?

   _____
   _____
   _____

Name _____ Class _____ Date _____

CHAPTER **3**
Investigation

Matter, Energy, Space, and Time

Conservation of Matter

Does the amount of matter (mass) in a substance remain the same during a change of phase?

PURPOSE

To compare the mass of a substance in the solid state to its mass in the liquid state

MATERIALS

balance
250-mL beakers (2)
graduate

water
uniformly sized ice cubes
glass stirring rod

PROCEDURE

1. Using the balance, determine the mass of an empty beaker. Record the mass in the Data Table.

2. Add 100 mL of water to the beaker, and determine the total mass of the beaker and the water. Record this mass in the Data Table.

3. To determine the mass of the water alone, subtract the mass of the empty beaker from the total mass of the beaker and the water. Record the result in the Data Table.

4. Set aside beaker 1 with the water. Determine the mass of the second beaker. Record this mass in the Data Table.

5. Add four or five ice cubes to this beaker, and determine the total mass of the beaker and the ice cubes. Record this mass in the Data Table.

6. Determine the mass of the ice cubes by subtracting the mass of beaker 2 from the total mass of the beaker and ice. Record this mass in the Data Table.

7. Add the ice cubes to beaker 1, the beaker containing the water, taking care not to spill any water or ice. Stir gently until all the ice cubes melt.

8. Carefully place beaker 1 with its contents on the balance, and determine the total mass of the beaker, water, and melted ice. Record this mass under the Data Table.

(continues)

HBJ material copyrighted under notice appearing earlier in this work.

Name _____ Class _____ Date _____

Chapter 3 Investigation (continued)

9. Add the mass of the water to the mass of the ice cubes to get the total initial mass.

10. Determine the total mass of the water and the melted ice by subtracting the mass of beaker 1 from the total mass of the beaker, water, and melted ice. This is the total final mass.

RESULTS

Data Table		
Mass of beaker 1	Mass of beaker 1 with water	Mass of water
Mass of beaker 2	Mass of beaker 2 with ice cubes	Mass of ice cubes

Total mass of beaker 1, water, and melted ice = _____

Total initial mass = _____

Total final mass = _____

CONCLUSIONS

1. How does the mass of water compare to the mass of ice?

   _____

2. What can you conclude from this investigation about the mass of a substance when the substance undergoes a phase change?

   _____
   _____
   _____

APPLICATIONS

A can of soda is placed in a freezer. After two hours it is removed. Has the mass changed? Why or why not? What physical characteristic of the soda has changed?

_____
_____

Name _____ Class _____ Date _____

CHAPTER **4** Investigation | **Elements, Mixtures, and Compounds**

Filtering a Mixture
Can sand and salt be separated from a mixture?

PURPOSE
To learn how to separate parts of a mixture by filtering

MATERIALS

graduate
sand (10 g)
salt (10 g)
ring stand
ring clamp
funnel
250-mL beakers (2)
filter paper

water
glass stirring rod
safety goggles
laboratory apron
evaporating dish
wire gauze
alcohol burner
tongs

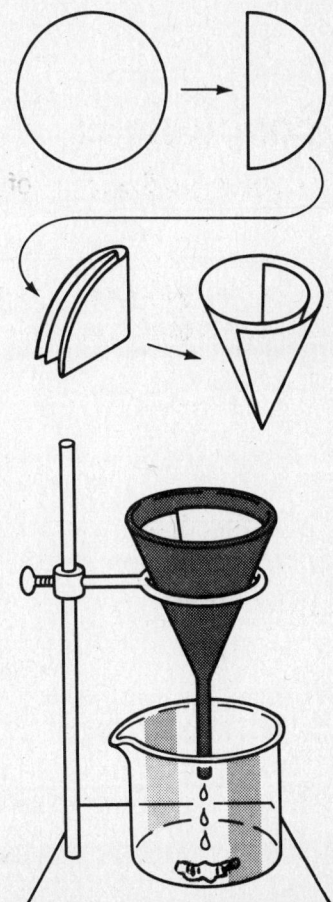

PROCEDURE

1. Prepare a mixture of 10 g of sand and 10 g of salt.

2. Set up the ring stand, funnel, and beaker as shown in the diagram.

3. Fold the filter paper in half and then in half again. Fit the filter paper into the funnel as shown in the diagram. One side of the filter paper should be a single layer and the other should be a triple layer. Wet the filter paper so it sticks to the funnel.

4. Make sure the funnel stem just touches the side of the clean beaker.

5. Place the mixture of sand and salt in a beaker. Add 100 mL of water to the mixture. Stir the mixture for five minutes.

6. Slowly pour the mixture of sand, salt, and water into the funnel. Do not pour the mixture above the top level of the filter paper.

7. After all the solution has passed through, remove the filter paper. Spread out the filter paper and allow it to dry. Leave the contents on the filter paper.

(continues)

Name _____ Class _____ Date _____

Chapter 4   Investigation   (continued)

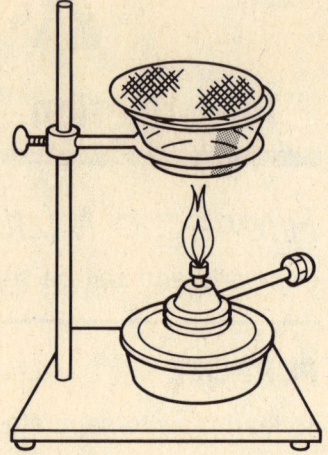

> **CAUTION:** Put on safety goggles and a laboratory apron, and leave them on for the remainder of this investigation. Alcohol is very flammable. Ask your teacher to instruct you in the proper use of the burner.

8. Prepare an evaporation setup as shown from the diagram. Make sure the evaporating dish is just above the burner flame.

9. Carefully pour half of the beaker's contents from the beaker into the evaporating dish. Do not let the contents boil over. Make sure that all water has evaporated before removing the evaporating dish. Turn off the burner. Remove the dish using the tongs.

RESULTS

1. What material was left on the filter paper?

   The sand.

   This material is called the *residue*.

2. What was the residue left in the evaporating dish?

   The salt

   How do you know?

   We observed

CONCLUSIONS

1. Why was the salt able to pass through the filter paper?

   It turn into small particles

2. Why was the sand not able to pass through?

   It does not desolve.

3. Could a mixture of salt and sugar be separated using this same technique? Why or why not?

   No sugar can not combine again therefor salt can.

(continues)

28

Name _____ Class _____ Date _____

Chapter 4 Investigation (continued)

4. What properties must a mixture have in order to be separated by filtering?

   Some a larger than others

5. What is the difference between a compound and a mixture?

   A compound you combine two things a mixture you mix a large amount of things.

APPLICATIONS

1. Describe other mixtures that could be separated by a filtering process.

   Air filter would keep dust a release clean air. Swimming pools use filter to clean the water.

2. How are filtration systems used in automobiles, sewage treatment facilities, and swimming pools?

   The sewage would be a clean substance after it pases through the filter.

Name _____ Class _____ Date _____

CHAPTER 5 Investigation | **Atoms and the Periodic Table**

Designing Electron Energy Shells

What is the electron arrangement in atoms of different elements?

PURPOSE

To build awareness of how electron energy levels occur

MATERIALS

electron-arrangement chart
thin, insulated copper wire
colored beads
fishing line

PROCEDURE

1. Select an element from the electron-arrangement chart.

2. Obtain enough insulated copper wire to make the electron levels for your element. Refer to the diagram for the proper diameter of each level. Allow enough extra wire for small knots to be tied on either side of the beads when they are placed on the wire.

3. Obtain the necessary number of beads for your element.

4. Make your first energy level. It should have two beads spaced opposite each other on the copper wire circle. To hold beads in their respective places, loop the wire into knots on either side of the beads. Then bend the wire into a circular or oval shape.

5. Repeat the process for each energy level of your selected element, using individual copper wire circles. The circles will be connected when all energy levels have been completed. Remember, no more than eight electrons can occupy an energy shell.

6. Use fishing line to connect the energy shells to one another so that they stay suspended and apart from one another. See the diagram. Save enough line to make a loop for hanging the model.

7. Suspend a label below the model, and hang the model from the ceiling. These models make an excellent display for studying electron energy levels.

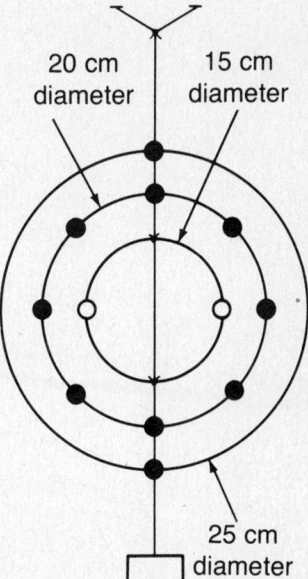

(continues)

Name _____ Class _____ Date _____

Chapter 5 Investigation (continued)

APPLICATIONS

1. How many electrons are in your element's outer energy shell?

   _____

   How many electrons are needed to complete the outer shell?

   _____

2. Would you classify your selected element as stable or unstable? Why?

   _____

   _____

3. Based on your response to the above question, describe how your selected element would react chemically.

   _____

4. With what other elements would it likely react?

   _____

5. Remember that atoms form compounds to fill outer energy shells with electrons. Suggest 2 or 3 compounds that your element might form with other elements? Why did you suggest these?

   _____

   _____

   _____

Name _____ Class _____ Date _____

CHAPTER 6 Investigation | Chemical Reactions

Baking Soda Reaction
What causes a chemical reaction to occur?

PURPOSE
To cause a chemical reaction that generates carbon dioxide

MATERIALS

safety goggles
laboratory apron
test tube
vinegar

measuring spoon
baking soda
balloon

PROCEDURE

CAUTION: Put on safety goggles and a laboratory apron, and leave them on throughout this investigation.

1. Blow the balloon up and release the air several times in order to stretch the balloon.
2. Fill the test tube about half full with vinegar.
3. Carefully add 1/4 teaspoon of baking soda to the test tube.
4. Quickly, while the solution is foaming, place a balloon on the test tube.
5. Repeat steps 1–3 several times. Sometimes it helps to shake the test tube a little.

RESULTS
Did a chemical reaction occur when you combined baking soda and vinegar? Explain your answer.

_____

_____

(continues)

Name _____ Class _____ Date _____

Chapter 6　Investigation　(continued)

CONCLUSIONS

Why did the gas expand the balloon?
_____
_____

APPLICATIONS

If a cake mix does not contain baking powder or baking soda, the cake will not rise when baked. Why?
_____
_____
_____

Name _____ Class _____ Date _____

CHAPTER 7
Investigation | Nuclear Reactions

Domino Reaction
How does a chain reaction occur?

PURPOSE
To demonstrate a model of how a single emission from a radioactive material can set off a chain reaction

MATERIALS
small square building block
dominoes (4 sets)

rulers (4)

PROCEDURE

1. Place a building block on a table or counter. Stand one domino upright in front of the square block and parallel to one of its sides. See the diagram.

2. Stand two more dominoes vertically, parallel, and symmetrically to the first domino. Continue this process until you have used all the dominoes, and a triangular shape is created as in the diagram.

3. Now, gently push the first domino away from the block so that it will fall over and hit the second group.

RESULTS

1. Describe the domino reaction that occurred.

_____

2. What happens to this reaction if there is not a symmetrical design?

_____

_____

3. Why does the domino reaction stop?

_____

_____

(continues)

HBJ material copyrighted under notice appearing earlier in this work.

35

Name _____ Class _____ Date _____

Chapter 7   Investigation   (continued)

CONCLUSIONS

1. Was your investigation successful? Why or why not?

   _____

   _____

2. Explain how and why the domino reaction occurred.

   _____

   _____

   _____

3. Describe the similarities between the domino reaction and a nuclear reaction.

   _____

   _____

APPLICATIONS

1. Repeat the domino reaction using four sides of the building block.

2. To demonstrate the similarities between the use of control rods in a nuclear reaction and rulers in a domino reaction, use a ruler to stop the reaction on one side of the square.

3. Now, reset the dominoes on four sides of the block, and using the rulers, try to stop the domino reaction on all four sides at the same time, halfway through the reaction. How many rulers and people will be needed to do this?

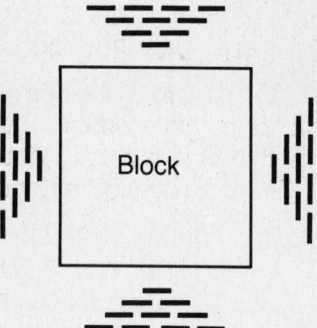

   _____

4. What would happen if the ruler were placed only halfway down a row of dominoes? How is the reaction affected?

   _____

   _____

5. Build a domino reaction using hexagonal or octagonal designs, and try to control the reaction. The designs of these reactions are in a two-dimensional plane. Try to imagine creating a three-dimensional chain reaction such as that which takes place in a real nuclear reaction.

Name _____ Class _____ Date _____

CHAPTER
8
Investigation
| **Chemical Technology**

Polymer Properties

What properties does a polymer exhibit?

PURPOSE

To explore the properties of a polymer

MATERIALS

safety goggles
laboratory apron
50-mL beakers (2)
250-mL beakers (4)
sugar
water
cornstarch
metal pie or cake pans (2)
mallet

PROCEDURE

CAUTION: Put on safety goggles and a laboratory apron, and keep them on throughout this investigation.

1. Make two solutions. In one beaker, mix enough sugar with water to make a thick, syrupy solution. In the second beaker, mix enough cornstarch and water to make a solution the consistency of a thick milkshake. There should be equal amounts of each solution.

2. This investigation will make a mess. Therefore, work in an open space that can be easily cleaned.

3. To test that each liquid exhibits normal characteristics of a liquid, pour each liquid into a pie pan. Do both take the shape of the container? Record your observations in RESULTS.

4. Now try the mallet test. Hit the liquid in the center of each pan as hard as you can with the mallet.

(continues)

Name _____ Class _____ Date _____

Chapter 8 Investigation (continued)

RESULTS

1. Do both liquids take the shape of the pans? What does this indicate?

   _____

2. What happened to the sugar/water liquid when struck with the mallet?

   _____

3. What happened to the cornstarch/water liquid when struck with the mallet?

   _____

4. What happened to the mallet when it hit the cornstarch liquid?

   _____

CONCLUSIONS

1. Why did the sugar/water liquid react as it did?

   _____

   What do you suppose happened to the molecules within this liquid?

   _____

2. Why did the cornstarch/water liquid react as it did?

   _____

   What do you suppose happened to the molecules within this liquid?

   _____

3. Which liquid exhibits properties of a polymer? Why?

   _____

   _____

4. Was your experiment successful? Why or why not?

   _____

   _____

   _____

(continues)

Name _____ Class _____ Date _____

Chapter 8　Investigation　(continued)

APPLICATIONS

1. Describe other properties that polymers may have.

   _____

   _____

   _____

2. List at least seven consumer products that are composed of polymers.

   _____

   _____

   _____

Name _____ Class _____ Date _____

CHAPTER 9
Investigation

Motion

Calculating the Speed of Sound

How can you determine the speed of sound through air?

PURPOSE

To calculate the speed of sound in air, using echoes

MATERIALS

large outside wall (gymnasium, etc.)
tape measure
tin-can drum
spoon or drumstick
watch with second hand, digital watch, or stopwatch

PROCEDURE

1. Walk away from a large outside brick wall in a straight line at right angles to that wall. Walk to a distance of 40 to 50 meters. Make a loud, sharp sound (such as hitting the bottom of a tin can with a spoon), and listen for the distinct echo. Move farther from the wall until you can hear the echo clearly.

2. Try tapping your "drum" several times in rapid succession, so that every new hit is timed to correspond with a returning echo. This will take some practice; you may need to adjust your distance from the wall. (When you do it right, the sounds of the echoes will disappear, since they are "covered up" by the striking of the can.) When you are successful, mark the spot on the ground and measure the distance to the wall. (Use a metric tape measure, or count measured strides if no tape measure is available.) Record the distance in the Data Table.

3. Now repeat the rhythmic tapping, again timing so that the "drum" is struck every time an echo returns. While doing this, have a partner count the number of beats you make in a ten second interval. Record this number in the Data Table.

4. Change your distance from the wall, and repeat your measurements of distance and number of beats required to "cover up" the echo.

(continues)

Name _____ Class _____ Date _____

Chapter 9 Investigation (continued)

RESULTS

	Data Table	
	Distance to Wall	Number of Beats in Ten Seconds
Trial 1		
Trial 2		

CONCLUSIONS

1. When you strike the "drum," how far does the sound travel before you hear it as an echo in trial 1? In trial 2?

   Trial 1 _____

   Trial 2 _____

2. If you know how many "beats" you make in ten seconds, figure out how many there are in one second.

   Trial 1 _____

   Trial 2 _____

3. If you know how many beats per second you made (from above), figure out how many seconds per beat there were. (This will be the reverse of the answers above.)

   Trial 1 _____

   Trial 2 _____

4. If you know how many seconds per beat there were, you also know the amount of time it took the sound to travel to the wall and return to you. You can use this to figure out the speed of sound. Just divide the distance the sound traveled (round trip) by the time it took to travel that distance.

   Trial 1 _____

   Trial 2 _____

APPLICATIONS

1. If you were in a Concorde supersonic jetliner, traveling over the Atlantic Ocean at 1.5 times the speed of sound (Mach 1.5), how far would you travel in one second?

   _____

Name _____ Class _____ Date _____

CHAPTER 10
Investigation

Work, Energy, and Machines

Defining Work

How can you determine the amount of work needed to perform several jobs?

PURPOSE

To figure out the amount of work required to complete some simple tasks

MATERIALS

classroom chair
string
spring scale (high capacity)
science textbook

PROCEDURE

1. Push a classroom chair just hard enough to make it slide a short distance across the floor. To calculate the amount of work required to move a chair in this manner, you must know the amount of force needed to slide the chair horizontally. You must also know the distance that you wish the chair to be moved.

2. To measure the force required to move the chair, tie a short piece of string to a leg of the chair. Attach the other end of the string to a high-capacity spring scale. Hold the spring scale parallel to the floor. Gently pull on the spring scale. Observe the force value as the chair moves at a slow, but constant, speed. (Be sure to keep the spring scale parallel to the floor while taking the reading.) Repeat this test several times, and record all of the trials in Data Table 1.

3. Now, imagine that you want to move the chair from where it is to the doorway of the classroom. Measure the straight-line distance to the doorway. Measure this distance; do not actually move the chair. Record this distance in Data Table 1.

4. Tie string around your science textbook so that you can pick it up with the spring scale. Lift the textbook slowly to a height of 50 cm above the desk. Record the force and the height in Data Table 2.

5. Leave the string on your textbook, and find the total weight of the textbook and string. Record the weight in Data Table 2.

(continues)

Name _____ Class _____ Date _____

Chapter 10 Investigation (continued)

6. Now use the first spring scale to drag your textbook 50 cm across a level desk. Record the force in Data Table 2.

RESULTS

Data Table 1				
	Trial 1	Trial 2	Trial 3	Average
Force required to slide chair (N)				
Distance chair is to be transported			_____ m	

Data Table 2	
Force needed to lift science textbook	_____ N
Distance textbook was lifted	_____ cm
Force needed to drag science textbook	_____ N
Distance textbook was pulled across desk	_____ cm
Weight of science textbook	_____ g

CONCLUSIONS

1. Find the average force required to slide the chair along the floor during your trials in part 1.

   _____

2. Calculate the work that would be required to slide your chair to the classroom door. (Work = force × distance)

   _____

(continues)

44   HBJ material copyrighted under notice appearing earlier in this work.

Name _____ Class _____ Date _____

Chapter 10 Investigation (continued)

3. Calculate the work required to lift your science textbook to a height of 50 cm.

   _____

4. Calculate the work required to drag your science textbook 50 cm.

   _____

5. Which task required more work, lifting your textbook 50 cm or dragging your textbook 50 cm?

   _____

6. When the textbook has been lifted 50 cm and is being suspended above the desk, what is its gravitational potential energy?
   (gravitational potential energy = weight × height)

   _____

APPLICATIONS

Why is it usually easier to slide rather than carry a heavy object?

_____
_____
_____
_____

CHAPTER 11 Investigation | Temperature and Heat

Phase Change of Water

How much heat is required to melt ice?

PURPOSE

To determine the heat of fusion for water and to compare it to the accepted value

MATERIALS

large container for ice
crushed ice
water
triple-beam balance
polystyrene foam cup
thermometer
thermometer clamp
ring stand
paper towels
glass stirring rod

PROCEDURE

1. Your teacher will prepare a large container of crushed ice to be used by the entire class. The ice should be in enough water to elevate its temperature to 0°C.

2. Using the triple-beam balance, determine the mass of a polystyrene foam cup. Fill the cup three-quarters full with tepid water. Determine the mass of the cup and water combined. Record both of these values in RESULTS.

3. Place the thermometer in its clamp, and attach it to the ring stand. Lower the thermometer into the cup, and record the initial temperature of the water.

4. Take several thicknesses of paper towel to the location of the ice water. Dip your fingers in, and remove several pieces of ice. Place the ice on the paper towel, and return to your station. Blot the ice dry; then place it in the cup containing the water. Carefully stir the water with a glass stirring rod. Do not spill any water, and be sure not to hit the thermometer bulb since it may break. When all of the ice has melted, be sure the water is thoroughly stirred. Record the lowest temperature reached by the water.

5. Raise the thermometer above the cup, and hold the stirring rod above the cup also. Catch any water drops in the cup; then place the cup on the balance and record its total mass (cup + water + melted ice).

(continues)

Name _____ Class _____ Date _____

Chapter 11 Investigation (continued)

RESULTS

1. Mass of the cup _____ g

2. Mass of cup + tepid water _____ g

3. Mass of cup + water + melted ice _____ C

4. Initial temperature of tepid water _____ C

5. Final temperature of cooled water _____ C

CONCLUSIONS

1. By subtraction, find the mass of the tepid water you used in this experiment.

2. Use subtraction to find the mass of the ice you added to the water.

3. Use subtraction again, this time to find the decrease in water temperature due to the ice melting.

4. Find the heat loss of the water needed to cause the ice to melt completely. Use the formula, $H = (m)(c)(T)$, from the textbook. This will represent both the heat loss in the water and also the heat gain in the melting ice. (Your answer will be in calories.)

5. Change the amount of heat from calories to Joules. (There are 4.2 Joules of heat in every calorie.)

6. Divide the heat gained in the melting of the ice by the mass of the ice that melted. This is your experimental value for the heat of fusion of water. (Be sure to calculate it in Joules/gram.)

(continues)

Name _____ Class _____ Date _____

Chapter 11 Investigation   (continued)

7. Is your experimental answer greater or less than the accepted value of 335 Joules/gram?

   _____

8. Which steps in the procedure do you think might account for any difference in the answers? Why?

   _____
   _____
   _____

9. Can you think of a way to increase the accuracy of this experiment?

   _____
   _____

APPLICATIONS

1. Why does the outside of an icy glass of water become wet?

   _____
   _____
   _____

2. Explain the heat transfers that occur when you make ice cubes at home. Start with the tap water in the tray that you place in the freezer compartment of your refrigerator, and trace the heat flow as far as you can.

   _____
   _____
   _____
   _____
   _____
   _____

Name _____ Class _____ Date _____

CHAPTER **12** Investigation | **Electricity and Magnetism**

Electromagnetism
How can you make a magnet turn off and on?

PURPOSE

To construct an electromagnet and to test its properties

MATERIALS

#22 insulated copper wire (1 m)
16-penny nail
magnetic compass
D-cell batteries (2)
box of paper clips
spring scale (0–2.5 N)
ring stand

PROCEDURE

1. Take about 1 m of thin, insulated, single-strand copper wire and wrap it tightly (in close coils) around a 16-penny nail. Start at the head of the nail, and leave about 10 cm of wire free (unwound) at each end. When you reach the point of the nail, you may go back and overlap the other coils, but be sure to continue wrapping in the same direction that you started. It is usually a helpful trick to wind about five to ten loops at a time and then to stop winding and force the loops tightly together before continuing.

2. When you have finished wrapping the nail, strip or scrape the insulation from the exposed, uncoiled ends of the wire.

3. Lay the nail on the desk, and place a compass near the point of the nail. Slowly slide the compass from the point toward the head of the nail, always keeping about the same distance from the nail. Describe your observations in RESULTS.

4. Hold a D-cell battery between the exposed wire ends, keeping each end pressed tightly against one terminal of the battery. Repeat the experiment with the compass, and record your observations.

5. Now reverse the connections of the D-cell terminals and again slide the compass from end to end. Compare the compass's behavior to that of the previous trial.

(continues)

HBJ material copyrighted under notice appearing earlier in this work.

Name _____ Class _____ Date _____

Chapter 12   Investigation   (continued)

6. With a D-cell battery connected to the wires, bring the tip of the nail into contact with a pile of paper clips. See how many you can lift from the surface of the desk, and record this in RESULTS. While suspending the paper clips above the desk, remove a wire from the cell terminal and observe the result.

7. Try the same experiment with two D-cell batteries connected together. Place the positive terminal of one cell against the negative terminal of the other. Connect one wire to the remaining positive terminal and the other wire to the remaining negative terminal and again dip the nail into the pile of clips. Record in RESULTS the number of clips you can lift using the two cells together.

8. To measure the lifting power of your magnet, suspend a spring scale from a ring stand, and bring the head end of your electromagnet up to the balance hook. While the magnet is operating and magnetically attached to the hook, pull down gently and record in RESULTS the force required to break the contact. Try this first with one then with two D-cell batteries and record in RESULTS the maximum force reading before the contact is broken.

9. Work the coil of wire loose enough to slide the nail, and gently remove the nail from the coil. Lay the empty coil on the desk, and place the compass near one end. Slowly slide the compass toward the other end of the coil, and record the behavior in RESULTS. Connect a cell to the coil, and repeat the movement of the compass. Record your results.

10. Unwrap 50 cm of wire and repeat steps 6 through 9.

RESULTS

1. Describe the behavior of the compass needle as it is moved from one end of the nail to the other when no D-cell batteries are connected. (Use sketches if necessary.)

_____

_____

_____

(continues)

Name _____ Class _____ Date _____

Chapter 12 Investigation (continued)

2. Describe the behavior of the compass needle as it is moved when a D-cell battery is connected. (Use sketches if necessary.)

   _____

   _____

   _____

3. Describe the behavior of the compass when the D-cell battery connections are reversed.

   _____

   _____

4. Record the number of paper clips lifted using one cell.

   _____

   _____

5. What happens when the electrical contact is broken?

   _____

6. Record the number of paper clips lifted using two cells.

   _____

7. Lifting force (N) using one cell: _____

   Lifting force (N) using two cells: _____

8. Describe the behavior of the compass when it is moved near the empty coil.

   _____

   _____

9. Describe the behavior of the compass when it is moved near the empty coil with the cell attached.

   _____

   _____

10. How was the lifting power of the magnet affected when 50 cm of wire were unwrapped from the nail?

    _____

    _____

(continues)

Name _____ Class _____ Date _____

Chapter 12   Investigation   (continued)

CONCLUSIONS

1. What effect does breaking the electrical contact have on an electromagnet?

   _____
   _____

2. What effect does reversing the electrical contacts have on an electromagnet?

   _____
   _____

3. What effect does increasing the number of electric cells have on an electromagnet?

   _____
   _____

4. What are some things you might do to increase the lifting power of an electromagnet?

   _____
   _____
   _____

APPLICATIONS

1. Name some devices that use electromagnets in their operation.

   _____
   _____

2. What are some advantages of electromagnets over permanent metal magnets?

   _____
   _____
   _____

Name _____ Class _____ Date _____

CHAPTER **13**
Investigation
Light and Sound

Observing Fields of View
How much can you see reflected in a plane mirror?

PURPOSE
To determine the field of view for a plane mirror as you change your viewing position

MATERIALS
three-finger clamp
small rectangular plane mirror
ring stand
paper
meter stick
masking tape

PROCEDURE

1. Clamp or attach a rectangular plane mirror to a ring stand so that the mirror's center is at your eye level when you stand on the floor looking toward the mirror. Be sure that the mirror is perpendicular to the table top and arranged with its shorter side vertical. Record the height of the mirror in Data Table 1.

2. Make two movable paper markers for a meter stick by folding a half sheet of paper several times into a band and then wrapping the paper band tightly around the stick and taping it to itself. (The markers should be movable by hand along the meter stick, but they should not slip or flop about on their own when the stick is moved.)

3. Stand in front of the mirror, 1 m away. Hold the meter stick in a vertical position, with the 50-cm mark next to your eye while looking into the mirror. Adjust the position of the markers so that they appear to be at the very top and bottom edges of the mirror when you look at the meter stick's image. (You may want a partner to help with the adjustment while you look into the mirror.) Record in Data Table 1 the position of the markers on the meter stick and your distance from the mirror.

4. Double your distance from the mirror, and repeat the observation of the markers at the top and bottom edges of the mirror. Be sure to record all the values.

5. Stand at half of the original distance from the mirror, and repeat the observation of the markers at the top and bottom edges of the mirror. Be sure to record all the values.

(continues)

Name _____ Class _____ Date _____

Chapter 13   Investigation   (continued)

6. Try another measured distance, of your own choice, from the mirror. Repeat the observations and record the values as you did before.

7. Now rotate the mirror so that its longer side is vertical but its center is still at eye level. Record the new height of the mirror in Data Table 2. Use the same four distances from the mirror as you used in steps 3 through 6. Observe and record the positions at which the markers must be placed in order to appear at the top and bottom edges of this longer mirror.

8. Without holding up the meter stick, return to each of the four distances from the mirror. Observe the image of the background that is visible in the mirror at each position, and describe how much you can see of the room. Record your observations in Data Table 3.

RESULTS

Data Table 1

Height of mirror: _____ cm

Viewing distance from mirror (m)	Position of top marker (cm)	Position of bottom marker (cm)
1. _____	_____	_____
2. _____	_____	_____
3. _____	_____	_____
4. _____	_____	_____

(continues)

Name _____ Class _____ Date _____

Chapter 13 Investigation (continued)

Data Table 2

Height of longer mirror: _____ cm

Viewing distance from mirror (m)	Position of top marker (cm)	Position of bottom marker (cm)
1. _____	_____	_____
2. _____	_____	_____
3. _____	_____	_____
4. _____	_____	_____

Data Table 3

Viewing distance from mirror (m)	Description of background image in longer mirror
1. _____	_____
2. _____	_____
3. _____	_____
4. _____	_____

CONCLUSIONS

1. Calculate the distance between the markers on the meter stick. Do this for all of the trials. Add a column to Table 1 and Table 2 for these answers.

2. How much does the visible length of the meter stick change when you double your distance from the mirror?

_____

_____

(continues)

Name _____ Class _____ Date _____

Chapter 13   Investigation   (continued)

3. How much does the visible length of the meter stick change when you halve your distance from the mirror?

   _____

   _____

4. Divide the average visible length of the meter stick from Table 1 by the height of the mirror from Table 1.

   _____

5. Divide the average visible length of the meter stick from Table 2 by the height of the mirror from Table 2.

   _____

6. How does the visible length of the meter stick compare to the size of the mirror?

   _____

7. How does the amount of visible background change as you change your distance from the mirror?

   _____

   _____

   _____

APPLICATIONS

1. Does walking farther from a clothing-store mirror allow you to see more of your image? Explain.

   _____

   _____

2. As you move farther from a clothing-store mirror, what image would noticeably change?

   _____

   _____

Name _____ Class _____ Date _____

CHAPTER **14**
Investigation | **Computer Technology and Robotics**

Remote Control

How might you perform a task on command from a remote position?

PURPOSE

To illustrate the concept of a robot

MATERIALS

electromagnet (constructed during Chapter 12 Investigation)
D-cell batteries (2)
wire leads (with alligator clips)
steel ball
ramp or track
ring stand
pulley
thread or string
miscellaneous materials that might be needed to construct the robot
mass set

PROCEDURE

1. Using any electrical and mechanical equipment and materials available, construct a device that will be able to perform a simple task from a distance of at least 1 m. The task is to lift a 100-gram object from the floor to a height of at least 90 cm and keep it there for at least 30 seconds.

2. The device must have *at least* one mechanical part (such as a pulley, lever, pendulum, or inclined plane) and one electrical part (such as an electromagnet, a switch, a light, or a battery).

3. Plan, construct, and test your device. When you are sure of its reliability, notify your teacher so that the successful operation can be verified. Describe and sketch your device in RESULTS.

(continues)

Name _____ Class _____ Date _____

Chapter 14 Investigation (continued)

RESULTS

Draw a labeled sketch of your device. Include a brief written description and list of parts.

CONCLUSIONS

Can your device be called a robot? Explain your answer.

_____
_____
_____
_____

APPLICATIONS

1. List several uses of robots.

_____
_____
_____

2. Why are robots used to do these jobs?

_____
_____
_____
_____

Name _____ Class _____ Date _____

CHAPTER **15** | **Planet Earth**
Investigation

The Changing of the Seasons

PURPOSE

To show that changing the position of the sun in the sky causes the seasons to change also.

MATERIALS

small globe
flashlights or light bulbs
meter sticks (2)

PROCEDURE

1. Designate one person to hold the globe. During this investigation, the globe will be moved in a complete circle around the light source. The light source should be rested on the top of a meter stick which is perpendicular to the floor.

2. Hold the globe, right side up, 2 m from the light source.

3. Since the axis of the earth is tilted about 23.5°, tilt the globe slightly toward the wall behind the light source. Throughout the remainder of this investigation, the North Pole of the globe must continue to tilt in the direction of this wall.

4. Adjust the height of the globe so that the direct rays of light fall between Florida and Cuba.

5. In RESULTS, under "Position A," draw a diagram of the globe and light source as they appear in this position. Label the earth's axis, the North and South Poles, and the equator. Make sure the tilt is accurate.

6. Move the globe in a counterclockwise direction around the light source to a point one quarter of the way around the circle. Make sure the North Pole is still tipped toward the same wall. Adjust the height of the globe so that the direct rays of the light source are falling on the equator.

7. In RESULTS, under "Position B," draw a diagram of the globe and light source as they appear in this position. Label the earth's axis, the North and South Poles, and the equator. Again, make sure the tilt is accurate.

(continues)

HBJ material copyrighted under notice appearing earlier in this work.

Name _____ Class _____ Date _____

Chapter 15   Investigation   (continued)

8. Move the globe in a counterclockwise direction around the light source to a point one half of the way around the circle. Make sure the North Pole is still tipped toward the same wall. Adjust the height of the globe so that the direct rays of light are falling on central Australia. Also move the globe another 20 cm closer to the light source.

9. In RESULTS, under "Position C," draw a diagram of the globe and light source as they appear in this position. Label the earth's axis, the North and South Poles, and the equator. Again, make sure the tilt is accurate.

10. Move the globe in a counterclockwise direction around the light source to a point three quarters of the way around the circle. Make sure the North Pole is still tipped toward the same wall. Adjust the height of the globe so that the direct rays of light are again falling on the equator. Move the globe 20 cm farther from the light source than it was at "Position C."

11. In RESULTS, under "Position D," draw a diagram of the globe and light source as they appear in this position. Label the earth's axis, the North and South Poles, and the equator. Again, make sure the tilt is accurate.

12. Move the globe in a counterclockwise direction around the light source to the original position. Make sure the North Pole is still tipped toward the same wall. Adjust the height of the globe so that the direct rays of light are again falling between Florida and Cuba. Move the globe another 20 cm farther from the light source. The positions of the globe and light source should be identical to the diagram under "Position A."

RESULTS

Position A

Position B

(continues)

Name _____ Class _____ Date _____

Chapter 15 Investigation (continued)

Position C

Position D

CONCLUSIONS

1. Direct rays of the sun account for warmer weather.

   a. Which position accounts for warm months in the United States? _____

   b. What season is this position in the Northern Hemisphere?
   _____

   c. What is the distance between the globe holder and the light source? _____

(continues)

Name _____ Class _____ Date _____

Chapter 15   Investigation   (continued)

2. Indirect rays of the sun account for cooler weather.

   a. Which position accounts for cooler months in the United States? _____

   b. What season is this position in the Northern Hemisphere? _____

   c. What is the distance between the globe holder and the light source? _____

3. What seasons do the two remaining positions represent? _____

4. In which season is the Northern Hemisphere closest to the sun? _____

5. In which season of the Northern Hemisphere is the earth farthest from the sun? _____

APPLICATIONS

1. If the earth's axis were no longer tilted, what would be the effect on seasons?

2. Which has more effect on the seasons: distance from the sun or tilt of the axis? Why?

3. Where would you have to live on the earth in order to feel the least effect of the earth's tilt? Why?

Name _____ Class _____ Date _____

CHAPTER **16** | **The Solar System**
Investigation

How Does a Planet Remain in Its Orbit?

PURPOSE
To demonstrate the principle of centripetal force

MATERIALS
2-cm eye screws (3)
rubber balls of different sizes and weights (3)
string (1 m)

PROCEDURE
1. Twist the eye screw into the rubber ball.
2. Insert the string through the eye screw. Tightly tie off the string.
3. Slowly, begin spinning the ball around your head. You will find that you do not need to swing the ball very hard to keep it spinning. Be very careful not to hit anyone during this investigation.
4. Repeat steps 1 and 2 using the other two balls.

RESULTS
1. Describe what force you felt as the ball was spinning around your head.

2. Did the size and weight of the ball affect what you felt as you spun it around your head? Explain your answer.

(continues)

Name _____ Class _____ Date _____

Chapter 16 Investigation (continued)

3. What would have happened if the string had broken as the ball was spinning around your head?

_____
_____

CONCLUSIONS

1. What in the solar system is represented by your hand holding the string?

_____

2. What force does the string represent?

_____
_____

3. How is centripetal force affected by the size of the planets?

_____
_____
_____

APPLICATIONS

How is centripetal force affecting each of the following?

1. a car going around a curve

_____
_____

2. a tornado

_____
_____

3. a whirlpool

_____
_____

Name _____ Class _____ Date _____

CHAPTER 17 Investigation | The Universe

Constellations

PURPOSE

To show the relationship between stars in a constellation

MATERIALS

star chart for Northern Hemisphere
construction paper (any color)
pencils
scissors
clear overhead transparency
overhead projector
black transparency marker
bulletin board covered with black paper
white crayon or grease pencil

PROCEDURE

1. Look at the star chart you have been given. Choose one constellation from the chart, and check with your teacher to make sure no one else has chosen that constellation.

2. Using the star chart, carefully sketch the positions of the stars in your chosen constellation. Carefully cut out the stars in your constellation.

3. Place the clear overhead transparency on the overhead projector, and place your constellation over the transparency.

4. Using the transparency marker, fill in the holes on the black paper so that the constellation appears as a series of black dots on the transparency.

5. Following your teacher's instructions, project your constellation on the bulletin board, so that it is in the proper position relative to the North Star.

6. Using the white crayon or grease pencil, draw the stars and constellation on the bulletin board.

(continues)

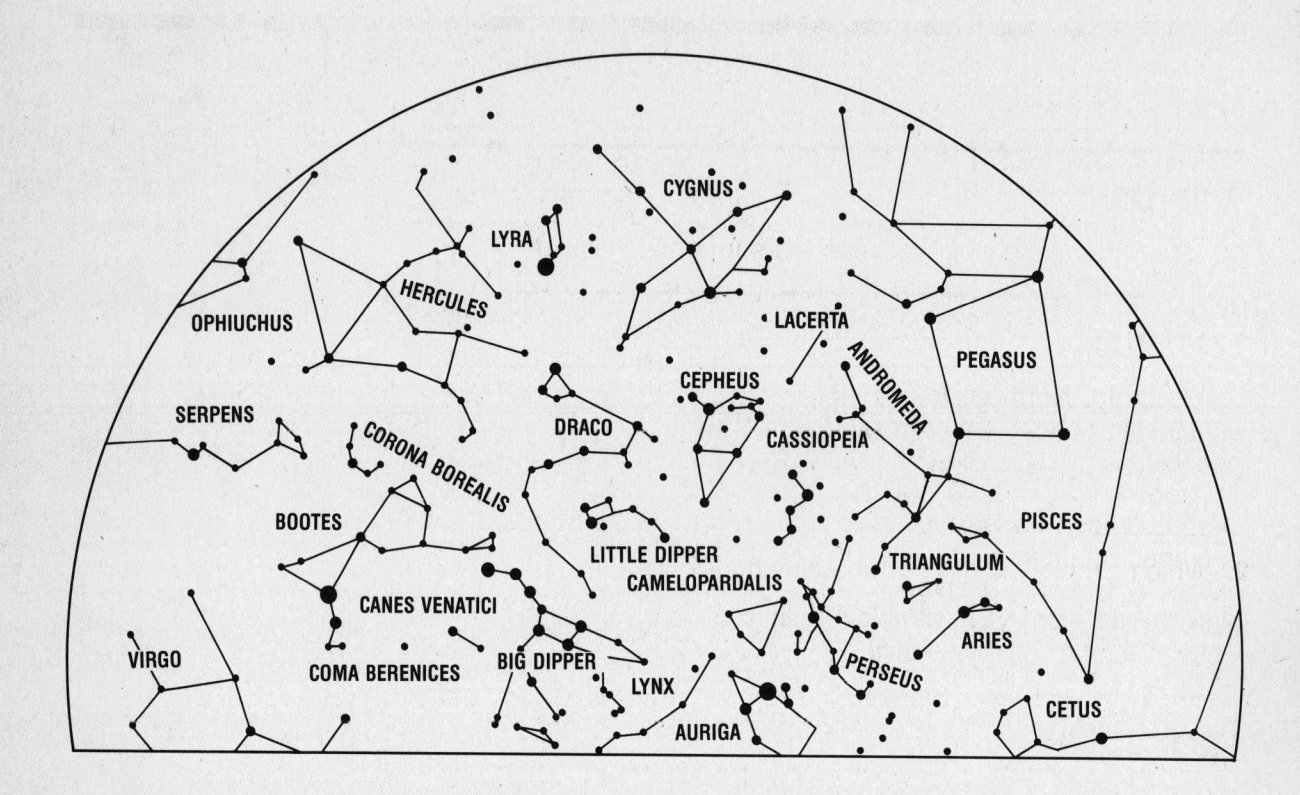

Chapter 17 Investigation (continued)

(continues)

Name _____ Class _____ Date _____

Chapter 17   Investigation   (continued)

RESULTS

Does the star chart on the bulletin board look like the star chart you were given at the beginning of the class? Why or why not?

_____
_____
_____

CONCLUSIONS

1. Which constellations resembled household objects?

_____
_____
_____

2. Which constellations resembled animals?

_____
_____
_____

APPLICATIONS

Do you see any other patterns in the star chart? Which constellations are involved and what would you name the "new" patterns?

_____
_____
_____

Name _____ Class _____ Date _____

CHAPTER 18 Investigation | Earth's History

Making a Fossil

PURPOSE

To produce a fossil artificially

MATERIALS

newspaper
dissecting trays or pieces
   of cardboard
clay
small bones or sea shells
   (clams or mussels)
safety goggles
laboratory apron
1000-mL beaker
water
paraffin wax or candles of
   different colors
250-mL beaker
crayon
tongs
hot plate

PROCEDURE

1. Cover your work area with newspaper.

2. Place the dissecting tray or the cardboard on the newspaper.

3. Flatten some clay in the tray until the clay is about 2 cm thick.

4. Gently press the bone or shell into the clay until a clear impression is made.

5. Remove the object from the clay.

   CAUTION: Put on goggles and a laboratory apron at this time and leave them on while using the water bath. Use a water bath carefully. Boiling water and hot beakers can cause serious burns.

6. Prepare a waterbath.

7. Fill the 1000-mL beaker about half full of water.

8. Place an amount of paraffin wax roughly equal in size to your "fossil" in the 250-mL beaker. Add a small portion of crayon to the wax for coloring as desired.

9. Using the tongs, carefully place the 250-mL beaker into the 1000-mL beaker.

(continues)

Name _____ Class _____ Date _____

Chapter 18 Investigation (continued)

10. Turn on the hot plate and allow the water to come to a boil (about ten minutes).
11. When the wax has melted, use the tongs to remove the 250-mL beaker from the water bath.
12. Carefully pour the melted wax into the fossil imprint in the clay.
13. When the wax has hardened, gently remove it from the imprint by pushing the clay away.
14. Use these same procedures to make several different fossils.

RESULTS

Arrange your fossils from smallest to largest and classify them according to type.

CONCLUSIONS

1. Carefully inspect the fossils. How do they compare in appearance to real fossils you have seen in photographs or in museums?

2. Under normal conditions in your classroom, do you think your imitation fossil will last a long time? Why or why not?

3. How might a similar process in nature preserve the remains of plants and animals?

APPLICATIONS

Does the absence of air in rocks have anything to do with how well a fossil is preserved? Explain your answers.

Name _____ Class _____ Date _____

CHAPTER 19 Investigation | The Changing Earth

Pangaea, the Supercontinent

PURPOSE

To identify current land masses that may at one time have been a part of a much larger land mass

MATERIALS

world map
scissors
black construction paper (2)

PROCEDURE

1. Using the world map on page 75, cut out North America, South America, Africa, Europe, Asia, and Australia. Place these land masses on the black construction paper.

2. Take your time and attempt to position the land masses so that they fit together. You will notice that some land masses will fit better than others, but you should be able to fit all the masses into one large mass.

RESULTS

1. Which land masses most obviously fit together?

   _____
   _____
   _____

2. Describe the positions of the various land masses in the large mass.

   _____
   _____
   _____
   _____

(continues)

Name _____ Class _____ Date _____

Chapter 19   Investigation   (continued)

CONCLUSIONS

1. Assuming that the land masses once fit together as you have shown, what forces could have moved them apart?

   _____
   _____
   _____
   _____

2. How might the positions of the continents be affected in the future?

   _____
   _____
   _____
   _____

APPLICATIONS

Research Continental Drift and describe some other evidence that the continents have moved around on the earth's surface.

_____
_____
_____
_____
_____
_____
_____
_____

(continues)

Name _____ Class _____ Date _____

Chapter 19   Investigation   (continued)

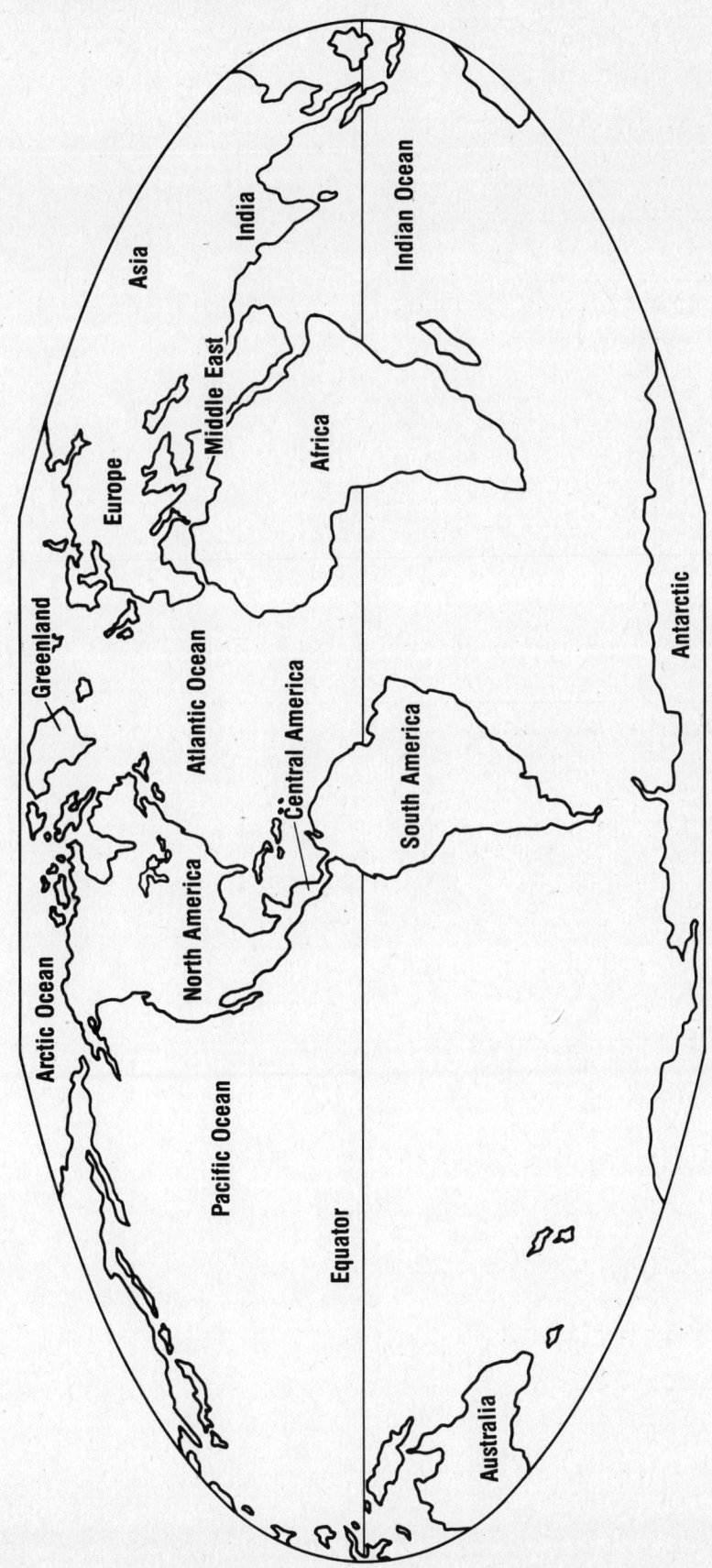

Name _____ Class _____ Date _____

CHAPTER
20
Investigation | **The Atmosphere**

Condensation of Water

PURPOSE

To show how condensation occurs in the atmosphere

MATERIALS

plastic wrap
small fish tank (1-gallon or 5-gallon tank will do)
rubber bands (large enough to wrap around the fish tank)
graduate
40-watt light bulb
75-watt light bulb
100-watt light bulb
socket and plug

PROCEDURE

1. Set up your investigation according to the diagram.
2. Turn on the light bulb as soon as your setup is completed.
3. Allow the light to shine on the setup for at least 30 minutes. Observe the space between the two pieces of plastic wrap.
4. When a large number of water droplets has formed, remove the upper layer of plastic wrap. Make sure that any water droplets sticking to the top layer of plastic wrap are carefully shaken onto the bottom layer.
5. The water droplets are all now lying on the top of the lower layer of plastic wrap.
6. Carefully pour the water (droplets) off the plastic wrap into the graduate. Record the amount of water in the Data Table.
7. Repeat the procedure two more times, first using the 75-watt bulb and then the 100-watt bulb. Each time record in the Data Table the amount of water produced.

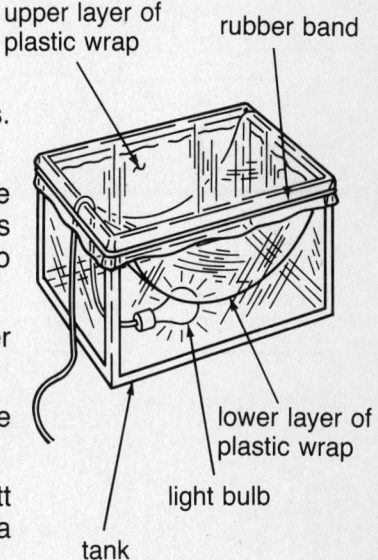

(continues)

Name _____ Class _____ Date _____

Chapter 20 Investigation (continued)

RESULTS

	Data Table
	Amount of water collected
40-watt bulb	
75-watt bulb	
100-watt bulb	

CONCLUSIONS

1. Why did you seal the top of the tank?

   _____

   _____

2. Why was the lower layer of plastic wrap used?

   _____

   _____

3. What factors caused the water to condense? Why?

   _____

   _____

APPLICATIONS

What could you have done to increase the rate of condensation?

_____

_____

_____

Name _____ Class _____ Date _____

CHAPTER
21
Investigation | **Weather and Climate**

Weathering

PURPOSE

To help explain "weathering" with simple math

MATERIALS

metric ruler
block of clay (4 cm × 4 cm × 4 cm)
kitchen knife

PROCEDURE

1. Measure each side of your block, making sure each side is 4 cm long.

2. Determine the area of each side of the block. Record these measurements in the Data Table.

3. Determine the total area of the block by adding the areas of the sides together. Record this measurement in the Data Table.

4. Cut the block in half along one side. Make sure the two blocks you now have created each measure 4 cm × 4 cm × 2 cm. Determine the area of each side of one of these blocks. Record these measurements in the Data Table.

5. Determine the total area of this block. Record this measurement in the Data Table.

6. Determine the total area of the two blocks. Record this measurement in the Data Table.

7. Cut each of the two blocks in half again. You should now have four blocks each measuring 4 cm × 2 cm × 2 cm. Determine the area of each side of one of these blocks. Record these measurements in the Data Table.

8. Determine the total area of one block. Record this measurement in the Data Table.

9. Determine the total area of all four blocks. Record this measurement in the Data Table.

(continues)

Name _____ Class _____ Date _____

Chapter 21 Investigation (continued)

RESULTS

Record your results in the Data Table.

Data Table

| | Area of each side of the block |||||| Total area of the block | Total area of all blocks |
	1	2	3	4	5	6		
4 cm × 4 cm × 4 cm block								
4 cm × 4 cm × 2 cm block								
4 cm × 2 cm × 2 cm block								

CONCLUSIONS

1. Suppose the original block of clay were a piece of clay or dirt in your back yard. How many sides could be exposed to the weather?

   _____

2. If "weathering" were to break this piece of clay in half, how many total sides would be exposed if you added all the sides of both clay pieces together?

   _____

3. As the clay block was cut into smaller and smaller pieces, how was the total area affected? How was the total mass affected?

   _____
   _____

4. In nature, as a rock is weathered and broken into smaller pieces, how is the rate of weathering affected? Why?

   _____
   _____
   _____
   _____

(continues)

Name _____ Class _____ Date _____

Chapter 21   Investigation   (continued)

APPLICATIONS

If you wanted to slow down part of a stream or brook, would you choose a lot of small rocks or a few large ones? Why?

_____

_____

_____

_____

Name _____ Class _____ Date _____

CHAPTER 22
Investigation | Ocean Currents

What causes ocean currents?

PURPOSE

To demonstrate ocean currents

MATERIALS

laboratory apron
revolving tray
pie pan
water
food coloring
dropper
stiff plastic the width of the stream table
stream table or large rectangular cake pan
silicon caulking
thermometer
stirring rod
several stones of varying sizes

PROCEDURE

CAUTION: Wear a laboratory apron throughout this investigation.

1. Place the revolving tray on a flat surface where it will be free to spin.

2. Place the pie pan on top of the tray and add water until the water is about 1 cm from the top of the pan.

3. Starting at the center of the pan and working toward one edge, place four or five drops of food coloring.

4. Slowly spin the tray in a counterclockwise direction, and observe the effect on the food coloring. Record your observations in RESULTS.

5. Stop the tray and remove the pan.

6. Empty the pie pan, place it back on the tray, and refill it with water to about 1 cm from the top of the pan.

7. Again, starting at the center and moving toward one edge, place four or five drops of food coloring.

8. Slowly spin the tray in a clockwise direction, and observe the effect. Record your observations in RESULTS.

(continues)

Name _____ Class _____ Date _____

Chapter 22 Investigation (continued)

9. Remove the pie pan from the tray and empty out the water.
10. Cut a piece of stiff plastic so that it fits snugly inside the stream table and across the width of the stream table. The stream table should be divided in half.
11. Seal the edges of the divider with caulking.
12. Pour cold water into one side of the stream table and hot water into the other side of the stream table. There should be at least a 20 degree difference in the temperatures of the water.
13. Add several drops of food coloring to the cold water and stir with the stirring rod.
14. Carefully remove the divider and observe what happens. Record your observations in RESULTS.
15. Empty the water from the stream table.
16. Set up the stream table exactly as before.
17. Add several stones to both sides of the stream table.
18. Carefully remove the divider and observe what happens. Record your observations in RESULTS.

RESULTS

1. What effect did a counterclockwise spin have on the food coloring?

   _____

   _____

2. Sketch your observations below.

(continues)

84

Name _____ Class _____ Date _____

Chapter 22 Investigation (continued)

3. What effect did the clockwise spin have on the food coloring?

   _____

   _____

4. Sketch your observations below.

5. Where did the greatest movement of food coloring occur—at the center or near the edges of the pan? Why?

   _____

   _____

6. When the divider was removed from the stream table the first time, how did the hot and cold water react?

   _____

   _____

7. When the divider was removed from the stream table the second time, how did the hot and cold water react? What effect did the rocks have on this movement?

   _____

   _____

CONCLUSIONS

1. What effect does the spinning pan demonstrate?

   _____

2. What does the counterclockwise spin demonstrate?

   _____

(continues)

Name _____ Class _____ Date _____

Chapter 22   Investigation   (continued)

3. What does the clockwise spin demonstrate?

   _____

4. On Earth, where does the greatest and least amount of particle movement occur in the oceans? How does this compare with the spinning pan?

   _____

   _____

   _____

5. What does the water movement in the stream table represent?

   _____

6. What do the stones represent?

   _____

7. On Earth, what causes currents to be deflected?

   _____

APPLICATIONS

1. How might wind affect surface currents?

   _____

   _____

2. How is the spread of water pollution linked to the coriolis effect and to temperature differences in the oceans?

   _____

   _____

   _____

Name _____ Class _____ Date _____

CHAPTER 23 Investigation | Living Things

A Classification System for Leaves

What are some of the characteristics by which leaves are classified?

PURPOSE

To show that by focusing on the differences among specific characteristics, it is possible to construct a classification system for leaves

MATERIALS

10 branches or stems (3–5 leaves attached on each)
poster board
cellophane tape
marking pen
notebook paper

PROCEDURE

1. Before beginning this investigation, obtain branches or stems from ten different plants. Each branch or stem should have three to five leaves attached.

2. Secure each stem to poster board with cellophane tape. Using a marker, label the samples *one* through *ten.*

3. Examine the stem and leaves on the card which you have labeled *one.* Determine if the leaves are simple or compound. A simple leaf is one that is not divided along the stem. Leaves from the oak, hibiscus, and rose are just a few that are classified as simple. Probably most of the leaves that you will examine will be simple in structure. A compound leaf is one that has two or more leaflets along a stem. A good example of a compound leaf is the leaf of the Royal Poinciana. After you determine the classification of leaf specimen number 1, enter either simple or compound beside *Leaf-1* under *Type* on the Data Table. All information gathered during this investigation is to be recorded on this Data Table.

(continues)

Chapter 23   Investigation   (continued)

4. Continue to examine the leaves labeled *one*. Determine how the leaves are arranged on the stem. They will be arranged either exactly opposite another leaf or alternately, with an empty space directly opposite each leaf. Once you have determined how the leaves are arranged on the stem, enter either opposite or alternate beside *Leaf-1* under *Arrangement of Leaves on Stem*.

5. Still examining the leaves labeled *one,* determine their shape. They can be classified as heart-shaped, spear-shaped, or round. Please note that while some leaves are quite easy to classify by shape, some will seem to fall between two categories. However, study the possibilities and make what you feel is the best selection and enter it beside *Leaf-1* under *Shape*.

6. Examine the outer edges of the leaves on card number 1. Are they smooth or do they have toothlike protrusions called serrations? Serrations can be very large or so small that they are detectable only by running your finger over the edge of the leaf. Decide in which classification the leaf belongs and enter smooth or serrated beside *Leaf-1* under *Edge*.

7. A leaf can also be classified by the system of veins on its internal structure. Leaves can have any one of four different systems of veins.
   Parallel—veins that run parallel to one another.
   Net—leaves that contain veins that look like a net.
   Pinnate—leaves that have one large vein running down their center with a series of smaller veins filling the remainder of these leaves.
   Palmate—leaves that have several large veins visible on their surface. Determine the vein structure of the leaves on card number one. Record parallel, net, pinnate, or palmate under *Vein Structure* beside *Leaf-1*.

8. Repeat all of the steps detailed above with the remaining nine leaf samples.

(continues)

Name _____ Class _____ Date _____

Chapter 23 Investigation (continued)

RESULTS

	Data Table				
	Classification of Leaves				
	Type	Arrangement of Leaves on Stem	Shape	Edge	Vein Structure
Leaf-1					
Leaf-2					
Leaf-3					
Leaf-4					
Leaf-5					
Leaf-6					
Leaf-7					
Leaf-8					
Leaf-9					
Leaf-10					

(continues)

Name _____ Class _____ Date _____

Chapter 23   Investigation   (continued)

CONCLUSIONS

1. Which characteristic do you feel was the most difficult for you to classify? Which was the easiest to classify? Why?

   _____
   _____
   _____
   _____

2. List the characteristic that occurred most frequently under each of the headings on Table 1. How would you describe a leaf with the most representative traits?

   _____
   _____
   _____
   _____

3. Can you suggest any other characteristics of the leaves that might also be used to aid in their classification?

   _____
   _____
   _____
   _____

APPLICATIONS

1. Using ten consecutive advertisements listed under used automobiles for sale in the classified section of a newspaper, construct and complete a classification system which includes at least five different characteristics common to all of the cars, such as make, model, country of origin, year, color, and so on.

2. Obtain a dichotomous key to leaves. Use the key to identify the type of plant from which each of your samples was obtained.

Name _____ Class _____ Date _____

CHAPTER 24 Investigation | Continuity of Life

Random Assortment of Genetic Traits

Why do children of the same parents have physical appearances that vary from one another?

PURPOSE

To show that during fertilization, the selection of physical traits is determined by random chance

MATERIALS

cardboard box (3)
blue poker chips (50)
white poker chips (50)

PROCEDURE

Independent assortment, now a basic law of genetics, was first proposed by Gregor Mendel. This law explains that the physical traits you have inherited from your parents are the result of a process called random selection. In other words, they are the result of pure chance. You received one half of the genes that determine your physical characteristics from your father and one half from your mother. The following steps of this investigation will allow you to examine the laws of genetics that govern independent assortment.

1. Mark the three cardboard boxes as "gene pool," "mother," and "father."

2. Place the 50 blue and the 50 white poker chips into the "gene pool" box. Shake the box or use your hand to stir the chips until they are thoroughly mixed.

3. Without looking, draw 50 chips from the box. Place these chips in the "father" box. Again, without looking, draw 50 chips from the "gene pool" box, and place these in the "mother" box. Examine the contents of the two boxes. Are there both blue and white chips in each box? (This step is representative of traits being selected from the gene pool at random in nature.)

(continues)

HBJ material copyrighted under notice appearing earlier in this work.

Name _____ Class _____ Date _____

Chapter 24 Investigation   (continued)

4. Once again, without looking, draw one chip at a time from the "mother" box and one at a time from the "father" box. Place them together on a table. Repeat this process until all of the chips have been drawn from each box. (This selection is representative of the fertilization process that occurs within species.) There are three possible combinations of chips now on the table. There can be two blue chips, one blue chip and one white chip, or two white chips.

5. The blue chips represent genes that are completely dominant for brown hair. The white chips represent genes that are completely recessive for blond hair.

RESULTS

1. What is the genotype of the father's haircolor? The mother's?

   _____

2. What are their hair color phenotypes? _____

   _____

3. How many pairs of blue chips, how many pairs of one blue chip and one white chip, and how many pairs of white chips were produced during the fertilization stage of your investigation?

   _____

   _____

   _____

   _____

4. Do a Punnett square for the parents' genotypes.

(continues)

Name _____ Class _____ Date _____

Chapter 24   Investigation   (continued)

5. Based on the Punnett square, how many pairs of blue chips should you have gotten? _____

   pairs of one blue and one white? _____

   pairs of white? _____

   How does this compare with your actual results?

   _____
   _____
   _____
   _____

CONCLUSIONS

1. When you chose the chips the first time, what were the chances that you would draw a chip of either color?

   _____
   _____
   _____
   _____

2. What is the only factor that determines which genes are contributed by each parent during reproduction?

   _____
   _____
   _____
   _____

3. What is the ratio of the number of dominant characteristics to the number of recessive characteristics yielded by your investigation? What does this tell you about an organism's chances of inheriting a trait that is dominant as opposed to inheriting a recessive trait?

   _____
   _____
   _____

(continues)

Name _____ Class _____ Date _____

Chapter 24 Investigation (continued)

APPLICATIONS

1. There are many diseases that may or may not be passed on to a member of the next generation because of the law of independent assortment. Suppose that a man has an inherited disease. The trait for this disease is dominant. Let this man be represented by one blue chip which carries the trait for the disease and one white chip which is the recessive gene that does not carry the disease. His wife, on the other hand, does not carry the disease. What is her genotype? _____

2. What are the chances this couple will pass this disease to their children? Support your prediction by constructing a Punnett square in the margin. _____
_____

3. If the couple had nine children in a row that inherited the trait for this disease, would it be possible for a tenth child to also inherit this dominant trait? Why?
_____
_____
_____
_____

Name _____ Class _____ Date _____

CHAPTER 25 Investigation | Evolution and Natural Selection

Variation Within a Species

Why have many species undergone physical changes over the duration of their existence?

PURPOSE

To show that variation of physical characteristics occurs within the same species

MATERIAL

meter stick or metric tape measure
notebook paper

PROCEDURE

Variation within a species is widely accepted as a component of the slow, ongoing process of evolution. These variations seem to account for the physical changes that occur within species over long periods. Charles Darwin's theory of natural selection proposes that those variations most beneficial to a species will be preserved through evolution.

1. In order to illustrate the variation among individual members of a species, carefully measure the height of each student in your class and record the results on notebook paper.
2. Transfer the data you collected to the Bar Graph.

RESULTS

Compare the information on the bar graph with your initial data in order to ensure that the graph is complete and accurate.

(continues)

Chapter 25 Investigation (continued)

Bar Graph: Variations in Student Heights

Number of Students in Each Height Range
25
24
23
22
21
20
19
18
17
16
15
14
13
12
11
10
9
8
7
6
5
4
3
2
1
0
137–143
144–150
151–157
158–165
166–173
174–180
181–188
189–195
196–203
204–210
Height Ranges (in cm)

(continues)

Name _____ Class _____ Date _____

Chapter 25   Investigation   (continued)

CONCLUSIONS

1. What height ranges contain the greatest number of students? Which ones contain the fewest?

   _____
   _____
   _____
   _____

2. Were the height ranges of students clustered toward the middle of the graph or were they fairly evenly distributed over its entirety?

   _____
   _____
   _____

3. Compute an average height for the students in the class. How does this average compare to the range with the greatest number of students?

   _____
   _____
   _____
   _____

4. If a class contained only males, how might this factor affect the results of the investigation?

   _____
   _____
   _____

(continues)

Name _____ Class _____ Date _____

Chapter 25 Investigation (continued)

APPLICATIONS

The first ancestors of humans were 1 or 1.3 meters in height and had brains approximately one half the size of those of modern humans. List some of the possible explanations as to how the gradual increase in stature and brain size enhanced the chances for survival of humans as a species.

_____
_____
_____
_____

Name _____ Class _____ Date _____

CHAPTER **26** | **Plants and Animals**
Investigation

Predator and Prey Relationships

What is the ideal relationship between predatory and prey species?

PURPOSE

To demonstrate that within a given environment there will usually be fewer predatory organisms than prey organisms

MATERIALS

50 white marbles
25 blue marbles
25 red marbles
empty oatmeal container
sock
scissors
notebook paper
graph paper

PROCEDURE

1. Place the 100 marbles in the oatmeal container. Stretch the sock over the opening of the cylinder, and cut off the toe. You should now be able to reach your hand into the container and randomly select the marbles.

2. Assign one person in your group to act as a recorder and one to act as a selector.

3. The white marbles represent sheep. Sheep are prey organisms. They serve as food for animals that are classified as predators. The blue marbles represent coyotes. The red marbles represent mountain lions. Coyotes and mountain lions are both predators.

4. Removes two marbles at a time from the container without looking.

5. The survival or death of the animals is based on the following formula. If two white marbles are drawn, representing two prey animals, both live. If one red or blue marble is drawn, representative of a predator, and one white marble is selected, the predator lives and the prey dies. If two red marbles, two blue marbles, or one red and one blue marble are drawn, both predators die.

(continues)

Name _____ Class _____ Date _____

Chapter 26   Investigation   (continued)

6. Assign another member of your group to collect the surviving organisms while another member of your group collects the ones that perish.

7. After all marbles are drawn, the recorder should note the number of surviving predators and surviving prey on a sheet of notebook paper. Label this information "first year."

8. The surviving organisms are then returned to their environment. (The marbles are placed into the container.)

9. Replacement organisms obtained through reproduction should be placed into the container by using the following formula. For each live predator add one white marble, two blue marbles, and one red marble. For each live prey add one white marble, one blue marble, and two red marbles.

10. The investigation should be done according to the process detailed above for a total of eight times. Each time record the number of surviving prey and predators. Label the information "second year," "third year," "fourth year," "fifth year," "sixth year," "seventh year," and "eighth year." This is representative of a predatory/prey relationship existing for an eight-year period.

RESULTS

Place the eight population cycles of the predators and prey on one graph for comparison.

CONCLUSIONS

1. What is the numerical relationship that exists between the populations of predators and prey?

   _____

   _____

   _____

   _____

2. Why is this relationship beneficial to both predator and prey?

   _____

   _____

   _____

   _____

(continues)

Name _____ Class _____ Date _____

Chapter 26 Investigation (continued)

3. If humans decreased the number of prey organisms available to the predators, what effect would there be on the populations of the predators and prey? How would you test your hypothesis?

   _____
   _____
   _____
   _____
   _____
   _____

4. List environmental variables that have not been accounted for in this investigation but that could possibly have an effect on the populations as they appear on your graph.

   _____
   _____
   _____
   _____
   _____

APPLICATIONS

Relate the findings of your predator/prey relationship to what might occur if this nation's farmers drastically underproduced a basic food crop.

_____
_____
_____
_____
_____

Name _____ Class _____ Date _____

CHAPTER
27
Investigation | **Human Biology**

Lung Capacity
How much air can your lungs hold?

PURPOSE
To measure the capacity of a human's lungs

MATERIALS
round balloon
meter stick
metric ruler
notebook paper

PROCEDURE

The capacity of a human's lungs may be expressed as either a measure of *tidal volume, expiratory reserve*, or *vital capacity*. The tidal volume refers to the amount of air that is taken in during normal breathing. The expiratory reserve is the amount of air that remains in the lungs after a normal breath is expelled. The vital capacity is the largest amount of air that can possibly be taken into and expelled from the lungs.

1. Blow up a round balloon several times until it becomes very flexible and is easy to fill with air.

2. In order to visualize your own tidal volume, take in a normal breath and exhale into the balloon only as much as as you do when you are breathing normally. Be careful to not let any of the air escape from the balloon.

3. Place the balloon next to an upright meter stick. Make a determination of its diameter by laying a ruler across the top of the balloon to a point where the ruler and the meter stick meet for form two 90-degree angles. Express the balloon's diameter in centimeters and enter the measurement on notebook paper under *Tidal Volume*.

4. Let all of the air out of the balloon.

5. Repeat the procedure nine more times. Enter each trial on notebook paper.

6. Breathe in and exhale only as much air as you do when you are breathing normally.

(continues)

Chapter 27 Investigation (continued)

7. Hold your breath and then forcefully expel as much air into the balloon as you can remove from the amount remaining in your lungs.

8. Measure the diameter of the balloon, and record it on notebook paper under *Expiratory Reserve.*

9. Repeat the procedure nine more times and record your results.

10. Take as deep a breath as you can. Exhale as much of the air as you possibly can into the balloon.

11. Measure the balloon's diameter. Record the measurement on notebook paper under *Vital Capacity.*

12. Repeat the procedure nine more times and record your results.

13. Compute an average diameter for the measurements recorded under *Tidal Volume, Expiratory Reserve,* and *Vital Capacity.*

14. Transfer the averages to the Bar Graph.

RESULTS

Compare the entries on your notebook paper to the information on the bar graph to ensure that it is complete and accurate.

(continues)

Name _____ Class _____ Date _____

Chapter 27 Investigation (continued)

Bar Graph: Lung Capacity

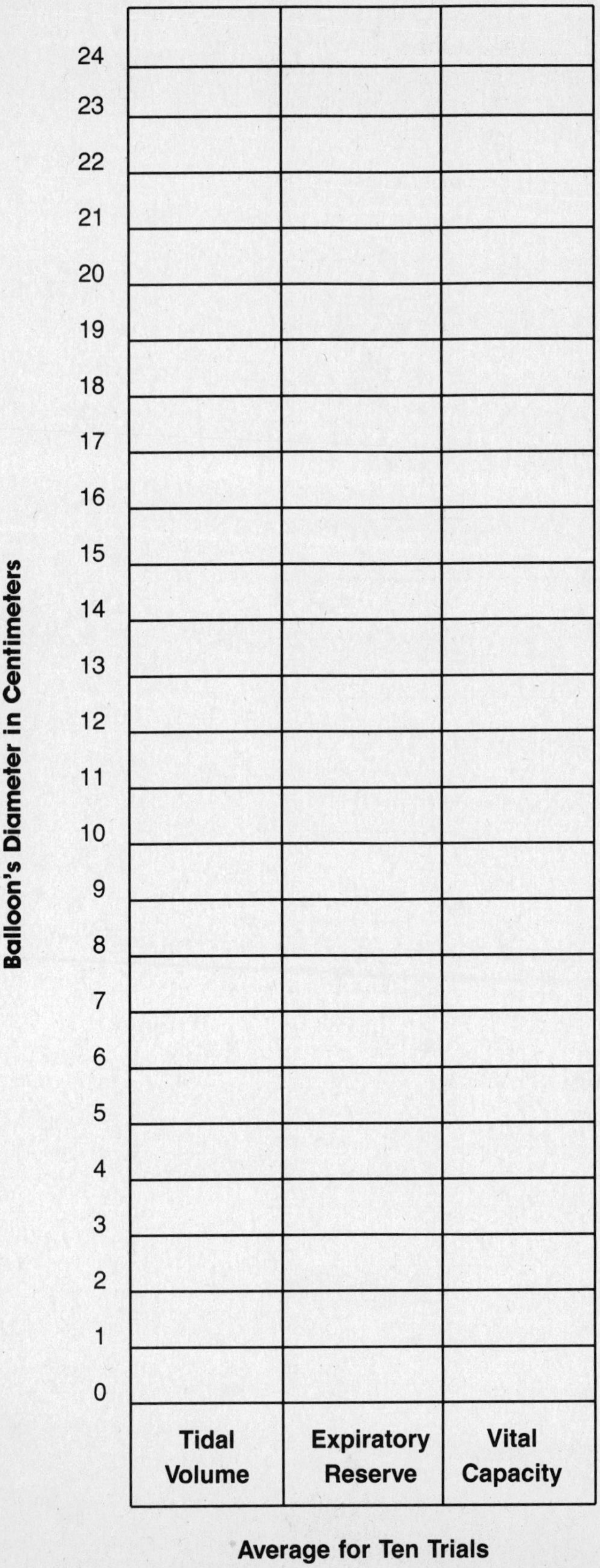

Balloon's Diameter in Centimeters (y-axis, 0–24)

Categories (x-axis): Tidal Volume | Expiratory Reserve | Vital Capacity

Average for Ten Trials

(continues)

Name _____ Class _____ Date _____

Chapter 27 Investigation (continued)

CONCLUSIONS

1. Given the fact that the larger the diameter of the balloon the greater its volume, which measurement of your lungs' capacity filled the balloon to its greatest volume? Which trials produced the smallest volume?

   _____

   _____

   _____

   _____

2. Which activity do you think would cause you to exhale the greater volume of air: sleeping or running? Why?

   _____

   _____

   _____

   _____

3. What factors would account for the differences in the averages obtained by individual students?

   _____

   _____

APPLICATIONS

What happens to the air you breathe after it enters your lungs?

_____

_____

_____

_____

Name _____ Class _____ Date _____

CHAPTER **28** | **Ecology**
Investigation

Ecological Treasure Hunt
What do you notice about your environment?

PURPOSE
To build awareness of specific features of the environment

MATERIALS
paper
pencil

PROCEDURE
Using natural features of your environment, describe in words or sketches the following:

1. something that is changing at a predictable rate
2. something that changes in minutes
3. something that changes over centuries
4. something soft
5. something hard
6. something shiny
7. something orange
8. a million of something
9. something fragrant
10. something man-made
11. something unique or rare
12. something that moves

(continues)

Name _____ Class _____ Date _____

Chapter 28   Investigation   (continued)

RESULTS

Share your ideas with your class. Where would you find at least six of the items in a similar ecosystem?

_____

_____

_____

_____

CONCLUSIONS

How would you rate your awareness of the environment?

_____

_____

_____

_____

Name _____ Class _____ Date _____

CHAPTER
29
Investigation

Managing Natural Resources

Corridor Problem Solving

How much land would be needed to build a highway and powerline corridor?

PURPOSE

To understand land-use planning

MATERIALS

paper
pencil
calculator

PROCEDURE

1. If there were to be a highway right-of-way of 200 m and it extended for 500 km, how many square kilometers of land would be needed? Hint: first convert to similar measurement.

2. Draw a sketch of the problem and label it. Try estimating; then calculate the exact distance. Record your answer in RESULTS.

3. If one-third of the highway was built through timberland and two-thirds through farmland, how many square kilometers of timberland would have to be cut?

   How many square kilometers of farmland would have to be paved?

RESULTS

1. Amount of land needed for step 1 of PROCEDURE.

   _____

2. Amount of timberland to be cut. _____

   Amount of farmland to be paved. _____

(continues)

Name _____ Class _____ Date _____

Chapter 29 Investigation (continued)

CONCLUSIONS

1. Describe the impact building this highway would have on forested lands in terms of

   plantlife: _____
   _____
   _____

   wildlife: _____
   _____
   _____

   watershed: _____
   _____
   _____

   water quality: _____
   _____
   _____

   aesthetics: _____
   _____
   _____

   commercial uses: _____
   _____
   _____

   recreational uses: _____
   _____
   _____

   pollution: _____
   _____
   _____

(continues)

Name _____ Class _____ Date _____

Chapter 29   Investigation   (continued)

2. What benefits would there be to building a four-lane highway across your state?

   _____

   _____

   _____

3. Evaluate the overall impact on natural resources versus need in constructing the highway. Should the highway be built or not? Explain your answer.

   _____

   _____

   _____

APPLICATIONS

1. Research laws relating to eminent domain.

2. Repeat the same process as above problem except for a powerline corridor. Assume a powerline were to be built going east/west (100 m wide and 300 km long) and another powerline north/south (100 m wide and 200 km long). How many square kilometers of land would be needed?

   _____

3. How many square kilometers would be needed to construct the highway and both powerline corridors?

   _____

Vocabulary & Concepts

Using a variety of formats, these worksheets (one per chapter) are designed to reinforce chapter vocabulary and concepts.

Vocabulary			Concepts		
Chapter	1	115	Chapter	1	116
Chapter	2	117	Chapter	2	118
Chapter	3	119	Chapter	3	120
Chapter	4	121	Chapter	4	122
Chapter	5	123	Chapter	5	124
Chapter	6	125	Chapter	6	126
Chapter	7	127	Chapter	7	128
Chapter	8	129	Chapter	8	130
Chapter	9	131	Chapter	9	132
Chapter	10	133	Chapter	10	134
Chapter	11	135	Chapter	11	136
Chapter	12	137	Chapter	12	138
Chapter	13	139	Chapter	13	140
Chapter	14	141	Chapter	14	142
Chapter	15	143	Chapter	15	144
Chapter	16	145	Chapter	16	146
Chapter	17	147	Chapter	17	148
Chapter	18	149	Chapter	18	150
Chapter	19	151	Chapter	19	152
Chapter	20	153	Chapter	20	154
Chapter	21	155	Chapter	21	156
Chapter	22	157	Chapter	22	158
Chapter	23	159	Chapter	23	160
Chapter	24	161	Chapter	24	162
Chapter	25	163	Chapter	25	164
Chapter	26	165	Chapter	26	166
Chapter	27	167	Chapter	27	168
Chapter	28	169	Chapter	28	170
Chapter	29	171	Chapter	29	172

Name _____ Class _____ Date _____

CHAPTER 1
Vocabulary
Science and Discovery

Complete the paragraph below about scientific methods by using vocabulary terms from the box. Not all terms will be used.

basic research	hypothesis	second
conclusion	meniscus	technology
data	science	theory
experiment	scientific method	variables

(1) _____ includes the body of knowledge that exists about the world. Many people work to make discoveries that contribute to this knowledge. (2) _____ is the name given to this kind of work. The knowledge is arrived at through the practice of (3) _____. The information that is learned is used in many ways. The use of scientific discoveries for practical benefits is called (4) _____. Scientists solve problems by following a series of steps. After stating a problem as a clear question, a scientist offers a (5) _____, a proposed answer. To test this proposal, the scientist conducts an (6) _____. The purpose is to test the (7) _____, the conditions that change according to the hypothesis. Two groups may be used in this type of experiment. One group is exposed to a changed condition, and a control group is not exposed to it. During the experiment, the scientist collects scientific facts called (8) _____. This information may be analyzed and evaluated before a (9) _____, a judgment based on the experiment, is drawn. The judgment may eventually become a (10) _____, a scientific explanation of known facts.

HBJ material copyrighted under notice appearing earlier in this work.

Name _____ Class _____ Date _____

CHAPTER 1 Concepts | **Science and Discovery**

A. In the space provided, indicate the units needed to measure the following quantities.

_____ 1. the distance from Philadelphia to Pittsburgh

_____ 2. the mass of a paper airplane

_____ 3. the volume of 20 drops of medicine

_____ 4. your body temperature

_____ 5. extremely cold temperatures occurring during a scientific experiment

B. In the space provided, indicate what a scientist would be measuring in each case.

_____ 1. A thermometer is placed in a beaker of liquid.

_____ 2. The length of a cage is multiplied by its width and height.

_____ 3. The distance from one building to the next is measured.

_____ 4. A clock is checked at the beginning and end of an experiment.

_____ 5. The length of a room is multiplied by its width.

C. In the space provided, indicate what part of a scientific method is being used.

1. John was given a dog for his birthday. He notices that the dog drinks the milk he gives it but does not eat much of the food. _____

2. John goes to the library and reads about dogs and dog food. Afterward he goes to the store to see what kind of dog food is available. _____

3. While at the store, John thinks the dog might eat liver but not chicken. _____

4. John buys both liver and chicken dog food and places them in separate dishes. _____

5. The dog eats neither. _____

6. John then thinks the dog might like beef. _____

7. John gives the dog beef-flavored dog food. _____

8. The dog eats the beef-flavored food. _____

Name _____ Class _____ Date _____

CHAPTER **2**
Vocabulary | **Science and Modern Technology**

Complete the crossword puzzle below using terms studied in this chapter.

Down

1. One diagnostic tool that uses both computers and X-ray technology is called a computerized _____ tomography (CAT) scanner.

3. Appliances have been improved by using the _____ circuit, an electrical unit containing many transistors.

4. A short name for an electrical unit containing many transistors is a _____.

5. Another name for advanced devices using electricity is _____ devices.

Across

2. A tiny device used to control electric current is called a _____.

6. Doctors can record the vibrations of body tissue by using nuclear _____ resonance (NMR) imaging.

7. Electronic devices that are used to store and process information in homes and offices are called _____.

8. The linking of a computer by telephone wires to other computers can form a computer _____.

Name _____ Class _____ Date _____

CHAPTER 2 Concepts | **Science and Modern Technology**

A. In the space provided, write the term that best completes each statement.

1. _____ control the speed or temperature of many appliances.

2. An improvement over the transistor is a series of transistors in a unit called an _____.

3. Devices that use electricity are called _____.

4. Information can be stored and used by _____.

5. High-definition television produces pictures that are _____ and _____ than those produced by regular television.

6. The check-out counter at the grocery store has a computer sensor that can read the _____ of each product that passes over it.

7. Cross-sectional pictures that show differences among tissues can be obtained using _____.

8. Instead of using scalpels, surgeons sometimes use _____ to cut through the skin.

9. _____ are used for communications, for weather forecasting, to study earth resources, and to study the solar system.

10. A _____ is an electronic device that can produce sounds that closely match the sounds of musical instruments.

B. In the space provided, answer the following question.

Summarize at least 3 ways in which modern technology has improved medical care, industry, and transportation.

_____

_____

_____

Name _____ Class _____ Date _____

CHAPTER **3**
Vocabulary | **Matter, Energy, Space, and Time**

A. Read each of the clues for terms related to matter, energy, space, or time. Then look across, down, and diagonally among the letters to find the hidden terms. Circle each term, and write it in the correct blank.

```
G P K S C H F N L U T
R H Q M D E N S I T Y
A L A D A A I O Q S G
V Q L E G T J R U O A
I W E I G H T S I L S
T B A L A N C E D I U
Y O T R T I M S R D S
```

1. Anything that takes up space and has mass is _____.

2. A _____ has a definite shape and volume.

3. A _____ has a definite volume but not a definite shape.

4. A _____ has neither a definite shape nor a definite volume.

5. Divide the mass of an object by its volume, and you get its _____.

6. The _____ of an object is caused by gravity acting on the object.

7. A _____ is an instrument used for measuring mass.

8. _____ is a force acting on objects with mass.

9. Energy transferred between objects at various temperatures is _____.

B. Write the letter of the term that best fits each description.

_____ 1. matter cannot be created or destroyed

_____ 2. going to a different state of matter

_____ 3. stored or inactive form of energy

_____ 4. energy cannot be created or destroyed

_____ 5. numbers used to locate an event

a. coordinates
b. gravity
c. kinetic energy
d. law of conservation of energy
e. law of conservation of mass
f. phase change
g. potential energy

HBJ material copyrighted under notice appearing earlier in this work.

119

Name _____ Class _____ Date _____

CHAPTER **3** Concepts | **Matter, Energy, Space, and Time**

A. In the space provided, write the state of matter for each substance.

1. _____ ice
2. _____ vapor
3. _____ rain
4. _____ star
5. _____ mercury
6. _____ putty
7. _____ diamond
8. _____ oxygen

B. In the space provided, write the term that best completes each statement.

1. On the surface of the earth, water can be found as a _____, a _____, and a _____.

2. A phase change can occur when the temperature is constant and the _____ is changed.

3. To decrease the time it takes to cook beets by boiling them, you could _____ the pressure in a pressure cooker.

4. If the volume of an object increases, its _____ decreases.

5. If you were to land on Saturn, your weight would _____.

C. What type of force is causing each action described below?

1. pen falling off a table _____
2. compass needle pointing north _____
3. an atomic nucleus being held together _____
4. an atomic nucleus breaking apart _____

120   HBJ material copyrighted under notice appearing earlier in this work.

Name _____ Class _____ Date _____

CHAPTER 5 Vocabulary | Atoms and the Periodic Table

A. Write the correct term in the spaces accompanying each definition. Then use the circled letters to spell the name of an important chart that lists all the elements.

1. relative mass, based on the atomic scale (_ _ _ _ ⓞ _ _ _ _ _)
2. smallest particle having an element's chemical properties (_ _ ⓞ _)
3. eight electrons in the outer shell of an atom (_ ⓞ _ _ _)
4. each row in the Periodic Table (_ _ _ _ _ ⓞ)
5. atoms of the same element that differ in mass (_ _ _ _ _ ⓞ _ _)
6. positively charged particles in an atom (_ ⓞ _ _ _ _ _)
7. lithium, sodium, cesium, potassium, and rubidium (_ _ _ _ _ ⓞ _ _ _ _ _ _ _)
8. an atom's central part, which houses protons and neutrons (_ _ _ _ ⓞ _ _)
9. electrically neutral particles inside the nucleus (_ ⓞ _ _ _ _ _ _)
10. negatively charged particles in an atom (_ _ _ _ ⓞ _ _ _ _)
11. fluorine, chlorine, bromine, and iodine (_ ⓞ _ _ _ _ _ _)
12. helium, neon, argon, krypton, and xenon (_ _ ⓞ _ _ _ _ _ _ _ _)
13. each column in the Periodic Table (_ _ _ _ ⓞ _)
14. the circled letters spell (_ _ _ _ _ _ _ _ _ _ _ _ _)

B. Write the definition for each of the following terms on the lines provided.

1. atomic mass units _____

2. atomic number _____

3. Periodic Table _____

4. mass number _____

5. subatomic particles _____

HBJ material copyrighted under notice appearing earlier in this work.   123

Name _____ Class _____ Date _____

CHAPTER 5
Concepts

Atoms and the Periodic Table

A. In each space provided, write the term that best completes the statement.

1. Thomson's experiments showed that particles have a _____ charge.

2. Rutherford reasoned that since atoms are neutral there must be a _____ charged particle, too.

3. If an atom has no net charge, the charge of the electrons must _____ the charge of the protons.

4. The atomic mass number is the total number of _____ and _____ inside an atom's nucleus.

5. An atom of oxygen can have _____ protons, _____ electrons, and _____ neutrons.

6. The _____ is the most stable state for an atom.

7. An electron close to the nucleus has _____ energy than one farther away.

8. When an electron _____ energy, it "jumps" to an orbit farther away from the nucleus.

9. A completely filled _____ has eight electrons in it and is very stable.

10. The valence-shell electrons determine how an _____ will react with another atom.

B. In the space provided, write the name of the group described.

1. good conductors _____

2. readily give up their one valence electron _____

3. dissolve in either water or alcohol _____

4. nonreactive _____

5. can be bent or flattened _____

6. have four or more electrons in outer shell _____

7. octet _____

124 HBJ material copyrighted under notice appearing earlier in this work.

Name _____ Class _____ Date _____

CHAPTER 6
Vocabulary

Chemical Reactions

A. Complete the sentences below about chemical reactions by using vocabulary terms from the box. Not all terms will be used.

acids	empirical	molecular
covalent bond	ion	molecule
electrolytes	ionization	synthesis reaction

1. A _____ is an element's or compound's smallest piece that can exist by itself.

2. An _____ is an electrically charged particle formed when an atom gains or loses valence electrons.

3. The process of gaining or losing an electron is called _____.

4. A chemical bond formed when atoms share valence electrons is called a _____.

5. Chemical formulas for ionic compounds are _____ formulas.

6. Ionic compounds that dissolve in water are called _____.

7. Compounds that release hydrogen ions in solution are _____.

B. Write the letter of the term that best fits each description.

_____ 1. produced when a transferring and sharing of valence electrons occurs

_____ 2. ionic compound containing a cation and an anion

_____ 3. short way of showing the composition of a compound

_____ 4. original substances that react together in a chemical reaction

_____ 5. substances produced by chemical reactions

_____ 6. fact that atoms are not destroyed in a chemical reaction

_____ 7. many involve the transfer of electrons from metal atoms to ions in solution

_____ 8. common between ions in solution

a. chemical bond
b. chemical equation
c. chemical formula
d. exchange reactions
e. law of conservation of mass
f. macromolecule
g. products
h. reactants
i. salt
j. substitution reactions

HBJ material copyrighted under notice appearing earlier in this work.

125

Name _____ Class _____ Date _____

CHAPTER **6**
Concepts

Chemical Reactions

A. In each space provided, write the term that best completes the statement.

1. Most atoms combine with other atoms to form _____.

2. A molecule of oxygen is more stable than two atoms of oxygen because it shares four _____ electrons.

3. When an atom loses valence electrons, it becomes a(n) _____ charged cation.

4. When an atom gains valence electrons, it becomes a(n) _____ charged anion.

5. An oxygen atom has six electrons in its outer shell. To become stable it must _____ electrons with another atom.

6. When magnesium transfers two electrons to oxygen, a(n) _____ bond is formed between the two atoms.

7. In molecular formulas, a subscript is written _____ the symbol for the element.

8. When two atoms of oxygen combine to form a molecule, its chemical formula is _____.

9. When salt is mixed with water, a(n) _____ is formed.

B. In the spaces provided, fill in the information needed to complete the chart.

Formula	Total Number of Atoms	Number of Indicated Atoms	Number of Indicated Atoms
N_2	_____	N _____	
P_4	_____	P _____	
NH_3	_____	N _____	H _____
CH_4	_____	C _____	H _____

126

Name _____ Class _____ Date _____

CHAPTER 7
Vocabulary | Nuclear Reactions

A. Complete the crossword puzzle below using terms studied in this chapter.

Down

1. type of nuclear reaction that combines nuclei
3. measures the rate of nuclear disintegration
4. the splitting of the nucleus
5. units used to measure the biological effect of exposure to radiation
6. a radioisotope that is used in experiments or treatments because its radioactivity makes it easy to locate

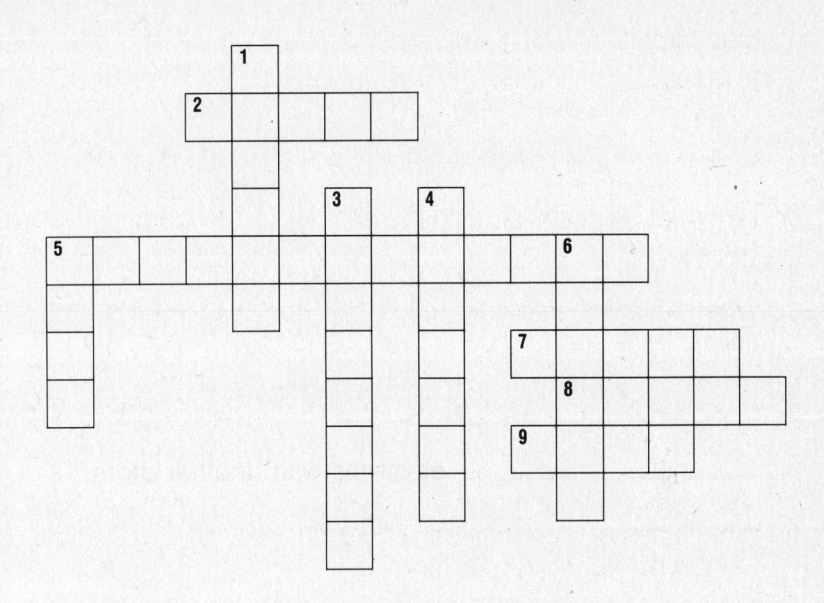

Across

2. equal to 37 billion nuclear disintegrations per second
5. invisible radiation
7. rays of pure energy
8. chambers used to show the tracks left by radiation
9. particles with a single negative charge

B. Write the definition for each of the following terms on the lines below.

1. alpha particle _____

2. background radiation _____

3. chain reaction _____

4. half-life _____

HBJ material copyrighted under notice appearing earlier in this work.

127

Name _____ Class _____ Date _____

CHAPTER 7
Concepts
Nuclear Reactions

Read each of the clues for terms related to nuclear reactions. Then look across, down, and diagonally among the letters to find the hidden terms. Circle each term, and write it in the correct blank.

```
O D L T F M U R A N I U M E G C
H P B X J N E R D Q T S Y Q A K
R V Z H P M A L P H A V C M I O
A H B G K L H N T U V T G U W B
B A C K G R O U N D S W U G L E
F L C F N E M W A C O O L A N T
E F D J I M I O Y F Z W B M X A
U L G F U S I O N X D R N M E S
Q I C K W I S P Y C K B P A D L
M F R A D I O I S O T O P E S F
I E O S E Q Z F O T H Z R N J B
A E Y A J A B X V N U C L E U S
```

_____ 1. type of radiation found in nature

_____ 2. measure of the biological effect of exposure to radiation

_____ 3. occurs when a reactor core reaches 3000°C

_____ 4. removes heat from a reactor core

_____ 5. time for one-half the atoms of a radioactive substance to decay

_____ 6. atomic nuclei combine

_____ 7. atomic nuclei split

_____ 8. high-speed electron particle

_____ 9. densest part of an atom

Name _____ Class _____ Date _____

CHAPTER 8
Vocabulary

Chemical Technology

A. Read each of the clues for terms related to chemical technology. Then look across, down, and diagonally among the letters to find the hidden terms. Circle each term, and write it in the correct blank.

```
S E C O A L T E M S F
I F S E I A M F L K O
C E T S R U B B E R T
C O E R S A L L O Y F
F O E S B T M I R A T
P O L Y M E R I E T O
O P P L A S T I C O T
```

1. _____ is a solid formed from the remains of plants.

2. A giant molecule made by chemically joining many identical smaller molecules is called a _____.

3. A _____ is a solid that can be molded when heated.

4. _____ is a polymer that is elastic.

5. _____ is any natural combination of minerals from which a metal may be separated or extracted.

6. _____ is the name given to a mixture of two or more metals or a mixture of a metal and a nonmetal.

B. Write the letter of the term that best fits each description.

_____ 1. a dark, oily, liquid mixture containing carbon and hydrogen
_____ 2. often found with petroleum
_____ 3. not made from natural plant or animal products
_____ 4. ultrathin tube of glass
_____ 5. any substance that helps plants grow

a. composite
b. cracking
c. fertilizer
d. natural gas
e. optical fiber
f. petroleum
g. synthetic fiber

129

Name _____ Class _____ Date _____

CHAPTER 8 Concepts
Chemical Technology

A. In the spaces provided, write the type of fossil fuel to which the clue refers. For some clues, you may need to write more than one answer.

1. made up of hydrocarbons: _____

2. when this fuel is burned, only carbon dioxide and water are given off: _____

3. a solid fuel that must be mined: _____

4. breaks down into many types of fuels: _____

5. when heated without air, forms three new substances: _____

6. a liquid and a gas that may be trapped together between layers of rock and must be drilled for: _____

B. In the spaces provided, write the name of the polymer to which the clue refers.

1. Long chains of ethylene molecules are used to form _____.

2. A polymer that will stretch, bend, and return to its original shape is _____.

3. Plastics such as _____ can be molded only once.

4. Plastics such as _____ can be repeatedly molded by heating and cooling.

C. In each space provided, write the term, phrase, or number that best completes each statement.

1. Bronze is a mixture of copper and tin; therefore, it is a(n) _____.

2. A 12-karat gold ring is _____ parts gold and _____ parts alloy.

3. Glass is made by heating quartz and sand and then _____.

4. Information can be passed from one place to another by using ultrathin _____.

Name _____ Class _____ Date _____

CHAPTER **9** | **Motion**
Vocabulary

A. Unscramble each term in column II, and write it in the blank to the left of its definition in column I.

I | II

1. _____: occurs when an object changes its position relative to its surroundings | 1. ntomio

2. _____: the length between two points | 2. tsaiecnd

3. _____: the background against which motion is observed and measured | 3. meraf fo efenrcree

4. _____: distance traveled per unit of time | 4. edpes

5. _____: distance traveled divided by the time it takes to cover that distance | 5. gaevrae desep

6. _____: speed in a particular direction | 6. tloveiyc

7. _____: rate at which the velocity of an object changes | 7. lcetnioreaca

8. _____: tendency of an object to resist any change in motion | 8. tieanri

B. Write the name of each description by using a vocabulary term from the box.

```
Law of Conservation of Momentum      Newton's Second Law of Motion
Uniform Circular Motion              Newton's Third Law of Motion
Newton's First Law of Motion
```

1. _____: For every action there is an equal but opposite reaction.

2. _____: An object at rest will remain at rest unless it is acted on by an outside force. An object in motion will continue to move in a straight line at a constant speed unless acted on by an outside force.

3. _____: When two or more objects collide, the total momentum of the objects is the same after the collision as it was before the collision.

4. _____: The acceleration of an object depends upon the net force acting on it and upon the mass of the object.

Name _____ Class _____ Date _____

CHAPTER 9 Concepts | Motion

A. In the space provided, indicate which one of Newton's Laws of Motion describes what is happening.

_____ 1. A jet plane is sitting at the gate while passengers are boarding.

_____ 2. The engines push out hot gases and move the jet down the runway.

_____ 3. While the plane is flying at 13,000 m, it hits an air pocket and drops 100 m.

_____ 4. A passenger who is not wearing a seat belt hits his head on the overhead storage compartment.

_____ 5. Another passenger is wearing her seat belt. She feels the force of the belt pushes down on her lap.

B. In the space provided, write the name of the type of motion described.

1. The moon is kept in orbit by the force of gravity. _____

2. In billiards, a cue ball strikes the eight ball and pushes it into the pocket.

   _____

3. A race car, starting up, reaches 120 km/hr in 6 seconds. _____

4. A motorcyclist is riding east on the interstate at 85 km/hr. _____

5. The pilot of a space shuttle lands on the runway and engages the brakes.

   _____

6. A rock breaks off the edge and drops into the Grand Canyon. _____

C. In the spaces provided, write the correct term for each clue.

1. a meter per second   _____

2. 50 km/hr due east   _____

3. 30 m/sec^2 _____

4. 5 kg × m/sec^2 _____

5. No force   _____

132 HBJ material copyrighted under notice appearing earlier in this work.

Name _____ Class _____ Date _____

CHAPTER 10 Vocabulary | Work, Energy, and Machines

A. Write the correct term in the spaces accompanying each definition. Then use the circled letters to spell a term meaning two or more simple machines working together.

1. a rod or bar that moves on a support point (_ _ _ _ _)
2. the rod or shaft part of a simple machine that works with a wheel (_ _ _ _ _)
3. a measure of how much of an energy input is available as useful energy output (_ _ _ _ _ _ _ _ _ _)
4. a force moving an object over a distance (_ _ _ _)
5. any tool that helps do work (_ _ _ _ _ _ _)
6. a wheel over which a rope or cord is passed (_ _ _ _ _ _)
7. a sloping surface that connects one level to another (_ _ _ _ _ _ _ _ _ _ _)
8. made up of one or two slanted faces or inclined planes (_ _ _ _ _)
9. does work in one movement (_ _ _ _ _ _ _ _ _ _ _ _)
10. the support point of a lever (_ _ _ _ _ _ _)
11. equal to a joule per second (_ _ _ _)
12. the ability to do work (_ _ _ _ _ _)
13. the energy of motion (_ _ _ _ _ _ _ _ _ _ _ _ _)
14. a machine made up of an inclined plane wound around a rod (_ _ _ _ _)
15. the circled letters spell (_ _ _ _ _ _ _ _ _ _ _ _ _ _)

B. Write the definition for each of the following terms on the lines below.

1. effort force _____
2. Law of Conservation of Energy _____
3. mechanical advantage _____
4. resistance force _____

HBJ material copyrighted under notice appearing earlier in this work.

Name _____ Class _____ Date _____

CHAPTER 10 Concepts | Work, Energy, and Machines

A. In the space provided, write the term or number that best completes each equation.

1. Force × distance = _____.

2. Gravitational potential energy = weight × _____.

3. Kinetic energy = 1/2 mass × _____.

4. One _____ = an apple picked up from the ground.

5. A joule per second is equal to a _____.

6. A kilowatt (kw) is _____ watts.

7. Work divided by time is equal to _____.

8. 746 W equals one _____.

B. In the spaces provided, answer the following questions.

1. When a quarterback throws a football, is work being done? Explain your answer. _____

2. As a girl throws a ball in the air, how does its potential energy change? _____

3. If a heavy ball and a light ball are dropped from a high-rise apartment, which ball will have more kinetic energy? Why? _____

4. According to the law of conservation of energy, what happens to kinetic energy as potential energy is lost? _____

5. As a moving object slows down because of friction, into what is its kinetic energy changed? _____

6. If a student pushes on a wall and the wall does not move, has any work been done? Why or why not? _____

Name _____ Class _____ Date _____

CHAPTER 11 Vocabulary | Temperature and Heat

Complete the sentences below about temperature and heat by using the vocabulary terms from the box. Not all terms will be used.

absolute zero	entropy	pascal
calorie	heat	radiation
Celsius scale	heat of fusion	temperature
conduction	insulator	thermal expansion
convection	internal energy	
cryogenics	Kelvin scale	

1. _____ measures the average kinetic energy of the molecules in an object or a substance.

2. Scientists often use the _____ to eliminate negative temperatures.

3. The point at which the molecules in a substance would have minimum kinetic energy is _____.

4. The study of matter at extremely low temperatures is _____.

5. The _____ is the metric unit of pressure.

6. _____ is the increase in the size of a substance due to an increase in temperature.

7. The energy that is transferred from one object to another because of a difference in temperature is called _____.

8. A _____ is the amount of heat needed to raise the temperature of one gram of water one Celsius degree.

9. Heat transfer when molecules collide with one another is _____.

10. A substance that is a poor conductor of heat is called an _____.

11. _____ is the transfer of heat by currents that move in liquids and gases.

12. _____ is a process of heat transfer in which energy goes from one place to another without passing through matter.

13. The _____ of a substance is the sum of the internal kinetic and potential energies of the substance.

HBJ material copyrighted under notice appearing earlier in this work.

Name _____ Class _____ Date _____

CHAPTER 11 Concepts | Temperature and Heat

In each space provided, write the term that best completes the statement.

1. The higher the temperature, the faster the _____ are moving.

2. The lower the temperature, the _____ the molecules are moving.

3. If the molecules of a gas increase their movement, then the _____ of the gas also increases.

4. The average kinetic energy of molecules is the _____ of the molecules.

5. Mercury and alcohol are used in a thermometer because they _____ and _____ more than the glass making up the thermometer.

6. A mercury thermometer works because mercury will _____ when heated and _____ when cooled.

7. A degree on the Celsius scale is 1/100 of the difference between the _____ and the _____ points of water.

8. When a solid is heated, it will turn into a _____ and then into a _____.

9. If oxygen molecules are heated to a high enough temperature, they will break up into _____.

10. At −273.15°C, molecules have very little _____.

11. When more air is added to a tire, the number of random collisions between air molecules and the tire _____.

12. One way to increase the number of collisions between molecules in a substance is to increase the _____ of the substance.

13. When a metal bar is heated, it will _____.

14. Cement sidewalks are made in blocks separated by joints so that they will not buckle in hot weather when they _____.

15. If the temperature of a gram of lead rises 1°C, that amount of heat is known as _____.

Name _____ Class _____ Date _____

CHAPTER 12
Vocabulary
Electricity and Magnetism

Complete the crossword puzzle below using terms studied in this chapter.

Down

1. the result of placing an iron core inside a coil of wire and running a current through the coil
3. a material that always has two poles
4. the unit used to measure electric current
5. the amount of charge in coulombs that passes a particular point in a conductor each second
6. a group of atoms that are coupled and lined up in the same direction
8. a pathway for electric current
11. a measure of how much a conductor holds back the movement of charge

Across

2. the unit of electrical resistance
7. a material whose electrical conductivity falls between that of conductors and insulators
9. the unit for measuring all electric charges
10. a material that does not conduct electricity
12. the unit for measuring the potential difference between two points
13. a material through which charges move easily

HBJ material copyrighted under notice appearing earlier in this work.

137

Name _____ Class _____ Date _____

CHAPTER 12 Concepts | Electricity and Magnetism

A. In each space provided, write the term that best completes the statement.

1. If you walk across a carpet on a dry day, your body may build up an _____.

2. When two positively charged or two negatively charged objects come together, they will _____ each other.

3. When a negatively charged object and a positively charged object come together, they will _____ each other.

4. If a piece of glass gains electrons, it becomes _____ charged.

5. If a piece of rubber loses electrons, it becomes _____ charged.

6. In atoms of metals, the valence electrons _____ from atom to atom.

7. Since it is electrons that move from atom to atom, negative charges are _____.

8. Plastic, glass, and rubber do not conduct electricity well because they have no _____.

9. Some electric charges are stationary, as in _____; others move through a wire to form a current.

10. Electric current is measured in _____.

11. When two ends of a wire are connected to a battery, electrons move from the _____ terminal through the wire to the _____ terminal.

12. A battery is a source of _____ that supplies the electrons and does work.

B. In the spaces provided, answer the following questions.

1. How do electrons flow when a battery is connected to a circuit? _____

2. When a circuit is connected to current produced by an electrical generator, how will the electrons flow? _____

138

Name _____ Class _____ Date _____

CHAPTER 13
Vocabulary

Light and Sound

A. Read each of the clues for terms related to light and sound. Then look across, down, and diagonally among the letters to find the hidden terms. Circle each term, and write it in the correct blank.

```
I R S T E F O A I P S T
P E T A D R M X R A Y S
S F T M H E I B E N O F
B R D P C Q C P L E N S
B A E L R U R I I O N M
S C R I L E O T B E T P
R T R T A N W C C E S L
A I T U S C A H E L L S
B O P D E Y V T E R O S
I N T E R F E R E N C E
```

1. From the top of a wave's crest to the bottom of a trough is the _____.

2. The intensity of a sound wave is measured in _____.

3. _____ is the number of waves that pass a particular point each second.

4. The interaction of waves in the same space is called _____.

5. Light waves from a _____ have only a single wavelength.

6. A _____ is a curved piece of glass or plastic that refracts light to form images.

B. Write the letter of the term that best fits each description.

_____ 1. can travel through different types of matter
_____ 2. can travel through matter and empty space
_____ 3. particles in the medium move parallel to the direction of the wave motion
_____ 4. distance between one crest and the next

a. diffraction
b. Doppler effect
c. electromagnetic wave
d. longitudinal wave
e. mechanical wave
f. wavelength

HBJ material copyrighted under notice appearing earlier in this work.   139

Name _____ Class _____ Date _____

CHAPTER **13** | **Light and Sound**
Concepts

A. In each space provided, write the term that best completes the statement.

1. When you shake a long rope tied to a chair, _____ moves down the rope, but the rope itself does not move along the path.

2. Sound waves are able to travel through three _____: solids, liquids, and gases.

3. _____, such as light, X rays, and radio waves can travel through matter and empty space.

4. If you shake a rope up and down, the wave travels in one direction and the particles move in a direction _____ to that direction.

5. If a slinky® is compressed and then released, a _____ is formed as the particles move in a direction parallel to that wave.

6. The amount of energy a wave has determines its _____.

7. As the wavelength gets longer, the frequency of the wave gets _____.

8. The shorter the wavelength, the _____ waves there are in a certain amount of time.

9. When the sound of a stereo is turned up, the _____ of the waves is increased.

10. If a gunshot was heard three seconds after it was fired 1,032 m away, the sound was traveling _____ m/s.

B. In the space provided, indicate the portion of the electromagnetic spectrum involved in each case.

1. may cause skin cancer _____

2. radar _____

3. given off by a hot pan _____

4. can be seen _____

5. used to catch speeders _____

6. can pass through solid matter _____

7. changed into electrical impulses by the eye _____

140

Name _____ Class _____ Date _____

CHAPTER **14** Vocabulary | **Computer Technology and Robotics**

A. Unscramble each term in column II, and write it in the blank to the left of its definition in column I.

I

1. _____: an electronic device that can store and process information

2. _____: physical components, or machinery, of the computer

3. _____: the information processing component of the computer

4. _____: the chip that contains the CPU

5. _____: set of programs for a computer

6. _____: "turn on/off" commands

7. _____: programs coded in what looks like everyday English

II

1. tcrpeomu

2. raehawdr

3. lrtneca gsipcornse tnui

4. srocimeosrporc

5. twrfeosa

6. nhecami agenugla

7. gihh-elelv galneausg

B. Fill in the blanks for the three main parts of a computer system by using the vocabulary terms from the box. One blank will not be filled.

arithmetic and logic section	main storage	control unit
printer	disk drive	tape machine
keyboard	video screen	

Input Devices **Central Processing Unit** **Output Devices**

_____   _____   _____
_____   _____   _____
_____   _____   _____

141

Name _____ Class _____ Date _____

CHAPTER **14** Concepts | **Computer Technology and Robotics**

Complete the crossword puzzle below using concepts studied in this chapter.

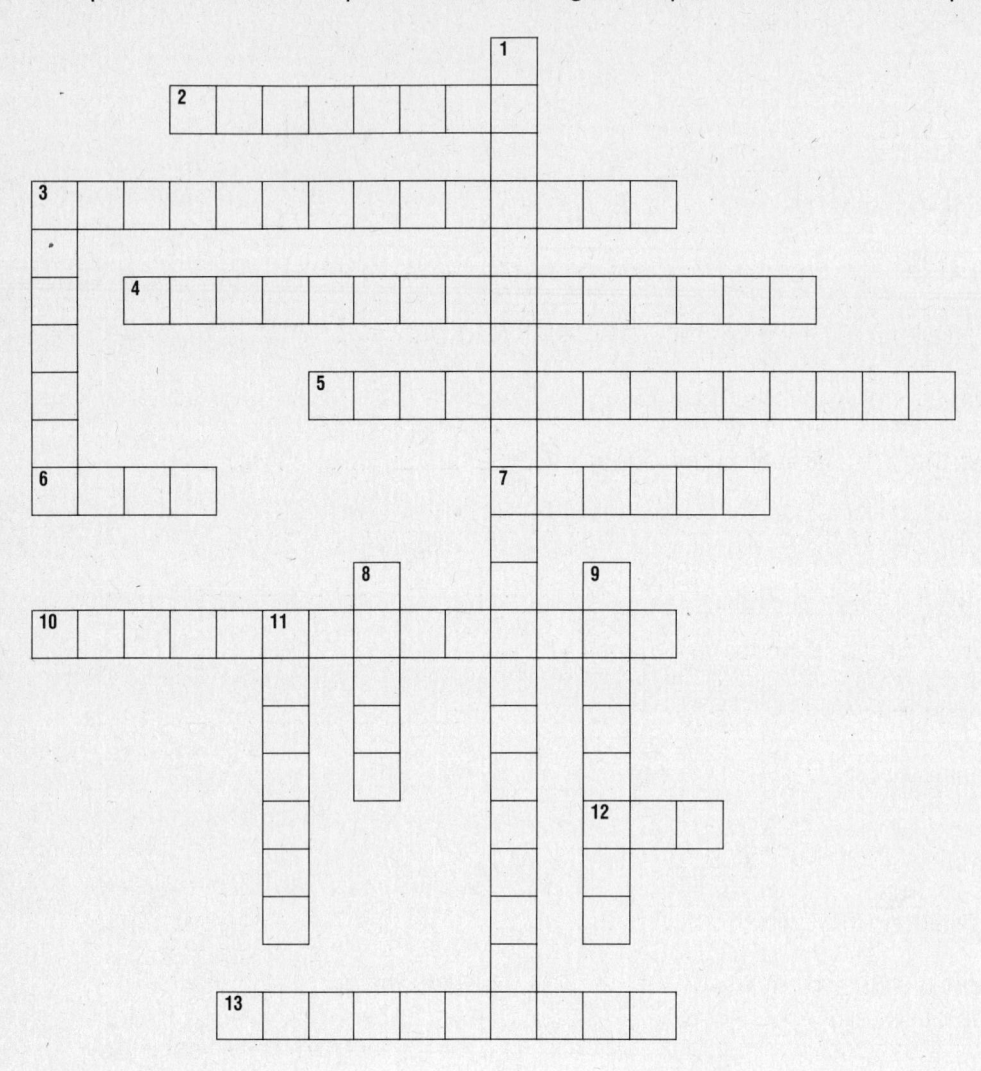

Down

1. the brains of a computer
3. instruments able to detect changes
8. computer with sensors
9. tells the computer when and what to do
11. computer drawing board

Across

2. machinery of a computer
3. very large, very fast machines
4. "turn on" and "turn off" commands
5. does what a typewriter used to do
6. height, width, and memory
7. gives information to the person running it
10. CPU or a chip
12. part of a computer that makes decisions
13. one step at a time

142

HBJ material copyrighted under notice appearing earlier in this work.

Name _____ Class _____ Date _____

CHAPTER **15** Vocabulary | **Planet Earth**

A. Write the correct term in the spaces accompanying each definition. Then use the circled letters to spell the name of the subject of this chapter.

1. the days when night and day are equal in length (_ _ _ _ ⊙ _ _ _ _)
2. locates the east-west position of a place (_ _ _ _ _ ⊙ _ _ _)
3. locates the north-south position of a place (⊙ _ _ _ _ _ _ _)
4. the two points at the top and bottom of the earth (⊙ _ _ _ _)
5. an imaginary line halfway between the two poles (_ _ _ ⊙ _ _ _)
6. extends halfway to the earth's surface (_ _ _ ⊙)
7. imaginary lines that are parallel to the equator (_ ⊙ _ _ _ _ _ _ _)
8. imaginary lines that run north and south (_ ⊙ _ _ _ _ _ _ _)
9. type of rock changed due to heat and pressure (_ _ _ _ _ _ _ _ ⊙ _ _)
10. the rocky layer of the earth above the mantle (_ ⊙ _ _ _)
11. noonday sun is highest or lowest from the horizon (_ _ _ _ ⊙ _ _ _ _)
12. the circled letters spell (_ _ _ _ _ _ _ _ _ _ _)

B. Write the letter of the term that best fits each description.

___E.___ 1. the rotation of the earth once in 24 hours
___H.___ 2. noonday sun directly overhead here during northern summer solstice
___G.___ 3. noonday sun directly overhead here during southern summer solstice
___B.___ 4. 24 hours of daylight here during northern summer solstice
___A.___ 5. 24 hours of darkness here during northern summer solstice
___F.___ 6. natural substances with specific chemical compositions

a. Antarctic Circle
b. Arctic Circle
c. crystals
d. igneous rock
e. mean solar day
f. minerals
g. Tropic of Cancer
h. Tropic of Capricorn

HBJ material copyrighted under notice appearing earlier in this work.

143

Name _____ Class _____ Date _____

CHAPTER **15** Concepts | **Planet Earth**

A. In each space provided, write the term that best completes the statement.

1. Atoms are made up of small subatomic particles: _____, _____, and _____.

2. Chemical compounds are made up of _____, always in the same arrangement.

3. Minerals are made up of similar _____.

4. Rocks are made up of different types of _____.

5. The crust of the earth is made up of different kinds of _____.

6. Rocks with a smooth, shiny, black, and glassy texture are _____.

7. Rocks with the imprint of shells in them are _____.

8. Rocks with bits and pieces of broken shells cemented together are _____.

9. Rocks with flattened black bands that are bent are _____.

10. Rocks with tiny crystals and large black crystals are _____.

B. Using the Mohs' Hardness Scale below, answer the following questions.

1.0 talc	6.0 orthoclase
2.0 gypsum	6.5 steel file
2.5 fingernail	7.0 quartz
3.0 calcite; copper penny	8.0 topaz
4.0 fluorite	9.0 corundum
5.0 apatite	10.0 diamond
5.5 glass	

1. Sample A will scratch a piece of glass but will not scratch a steel file. _____

2. Samples B & C can be scratched by your fingernail. Sample B will leave a scratch in Sample C. _____

3. Sample D will scratch everything listed in the box. _____

Name _____ Class _____ Date _____

CHAPTER 16 Vocabulary | The Solar System

Complete the sentences below about the solar system by using the vocabulary terms from the box. Not all terms will be used.

asteroids	ecliptic	meteors
astronomical unit	granules	outer planets
chromosphere	impact craters	photosphere
comets	inner planets	radiative layer
convective layer	light-year	solar prominences
core	maria	solar wind
corona	meteorite	sunspots

1. Mercury, Venus, Earth, and Mars are called the _____.

2. Jupiter, Saturn, Uranus, Neptune, and Pluto are the _____.

3. The plane of revolution used by eight planets is the _____.

4. One _____ is equal to the average distance of the earth from the sun—149,597,871 km.

5. At the sun's center is its _____.

6. The _____ of the sun extends to more than three-fourths of the way to the surface.

7. In the _____ of the sun, hotter matter from below rises, while cooler matter from above sinks.

8. The part of the sun that is visible from Earth is the _____.

9. The sun's 2500-km-thick layer of low-density gases is called the _____.

10. The sun's layer of very thin gases is called the _____, or "crown."

11. _____ are giant areas of hot gases created by very strong magnetic fields.

12. _____ are streams of hot gases that form in the corona and arc downward toward the surface of the sun.

13. _____ is a stream of particles that have escaped from the sun.

14. The broad plains of the moon are called _____.

15. _____ are depressions caused by collisions with asteroids or meteors.

HBJ material copyrighted under notice appearing earlier in this work.

Name _____ Class _____ Date _____

CHAPTER **16** Concepts | **The Solar System**

A. Use the diagram of the sun's layers to answer the questions that follow it.

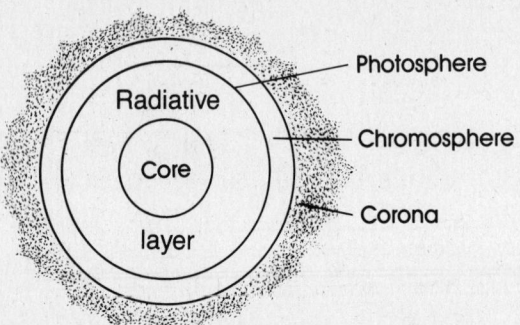

1. In which layer will the gravitational force be the greatest? _____ In which layer will it be the smallest? _____

2. Where will the density be the greatest? _____ Where will the density be the lowest? _____

3. Where will the atoms be closest together? _____ Where will the atoms be farthest apart? _____

4. Which layers are the darkest? _____ In which layer is light created? _____

5. In which layer do atoms glow red? _____

B. Match the following members of the solar system with their characteristics. There may be more than one answer for each statement.

    a. Mercury and Mars   **d.** Jupiter, Saturn, Uranus, and Neptune
    b. Venus and Earth   **e.** sun
    c. asteroids and comets

_____ 1. have a rock-like surface

_____ 2. contains most of the solar system's mass

_____ 3. have a very thick atmosphere of light gases

_____ 4. have the highest density of the planets

_____ 5. are the remains of debris from the solar system's formation

Name _____ Class _____ Date _____

CHAPTER 17
Vocabulary | The Universe

Complete the crossword puzzle below using terms studied in this chapter.

Down

1. A pair of stars revolving around each other is called a _____ system.

2. A _____ is a group of billions of stars.

3. Hipparchus grouped stars into six classes of brightness, called _____.

4. A collapsed star from which no light can escape is called a black _____.

7. A _____, which means "new" star, occurs when a star flares up and becomes up to ten thousand times brighter, as its hydrogen changes to helium.

Across

5. _____, or groups of stars, seem to form figures of animals and people.

6. As matter falls into neutron stars, a pulse of radiation is given off; these stars are called _____.

7. A _____ is a gas cloud.

8. A shrinking star shining with a weak, white light is called a white _____.

9. A _____ giant is a star that has used up most of its hydrogen and has expanded.

10. The twelve constellations found along the equator form the _____.

Name _____ Class _____ Date _____

CHAPTER 17 Concepts | The Universe

A. Numbering from one to five, place the life stages of a star in the proper sequence.

_____ 1. star becomes stable and gives off light for billions of years

_____ 2. loose cloud of gas and dust

_____ 3. runs out of hydrogen, the core contracts, and the other layers expand

_____ 4. star collapses and gives off a faint white light

_____ 5. gravitation squeezes the atoms so tightly that they heat up and a nuclear reaction begins

B. In the space provided, write the term from column II that best fits the description in column I.

I

_____ 1. results when a massive star collapses

_____ 2. star that undergoes a giant explosion and gives off a brilliant light

_____ 3. star that suddenly explodes, throws out gases into surrounding space, and then returns to normal size

_____ 4. very distant objects that give off both radio waves and huge amounts of energy

_____ 5. huge explosion that resulted in the Universe

II

a. nova
b. quasars
c. pulsar
d. black hole
e. Big Bang
f. nebula
g. supernova

C. In the space provided to the left of each statement, indicate which of the following instruments you would use.

a. Optical Telescope
b. Radio Telescope

_____ 1. You are observing the light changes in a variable star.

_____ 2. You are studying a quasar.

_____ 3. You are investigating an object believed to be a black hole.

_____ 4. You are observing a supernova.

Name _____ Class _____ Date _____

CHAPTER **18** Vocabulary | **Earth's History**

A. Complete the sentences below using vocabulary terms from the box. Not all terms will be used.

eras	fossil	interglacial age
evolution	geologic	Pangaea
extinct	ice ages	periods

1. The time interval from the formation of the earth to the present is called _____ time.

2. The time interval since the formation of the earth is divided into large units called _____.

3. The large units of time are divided into smaller units called _____.

4. Plants or animals that have died out are said to be _____.

5. _____ is the change in living things over time.

B. Put the eras from box I in chronological order beginning with the oldest era. List the items from box II beside the era in which they occurred. You will use some items more than once.

Box I

| Mesozoic | Precambrian | Cenozoic | Paleozoic |

Box II

a. ice age occurred
b. first humans
c. longest era
d. few fossil remains
e. primates evolved
f. first trees
g. first mollusks and fish
h. many trilobites
i. Age of Reptiles
j. Atlantic Ocean formed
k. first amphibians
l. dinosaurs evolved, then died

1. _____ Era _____
2. _____ Era _____
3. _____ Era _____
4. _____ Era _____

CHAPTER 18 Concepts | Earth's History

A. In each space provided, write the term that best completes the statement.

1. When the earth was first forming, heavier metals settled to the center because they were _____.

2. The minerals that were the _____ floated on the surface to form the crust.

3. The minerals that make up the mantle and crust are the _____.

4. The surface of the earth has changed due to the action of _____.

5. The earth's surface turned reddish when _____ combined with the iron minerals.

6. The water in the early ocean was hot and _____.

B. In the spaces provided, list the animals and plants that follow in the order of their development. Give a reason for your choice or a characteristic that makes an animal or plant group more advanced than the living things that existed before it.

      a. amphibians   **e.** fish with scales   **h.** primates
      b. blue-green algae   **f.** mammals   **i.** reptiles
      c. bony-plated fish   **g.** mollusks   **j.** trilobites
      d. worms

1. _____
2. _____
3. _____
4. _____
5. _____
6. _____
7. _____
8. _____
9. _____
10. _____

Name _____ Class _____ Date _____

CHAPTER 19
Vocabulary

The Changing Earth

A. Unscramble each term in column II, and write it in the blank to the left of its definition in column I.

I

1. _____: a violent shaking of the ground

2. _____: the place where the rocks break deep below the earth's surface during an earthquake

3. _____: the place on the earth's surface directly above the rock breakup during an earthquake

4. _____: the deep cracks caused by the earth's crust breaking into long blocks

5. _____: the rolling motion of the ground during an earthquake

II

1. hkrateuaeq

2. scuof

3. tneripece

4. stluaf

5. cimsies sevaw

B. Complete the sentences below using vocabulary terms from the box. Not all terms will be used.

asthenosphere	lithosphere	Richter
creep meter	Panthalassa	strain meter
laser beam	P waves	S waves

1. The _____ scale measures earthquake intensity.

2. Two instruments used to measure changes in distance between two points are the _____ and the _____.

3. _____ was the single ocean that existed long ago.

4. The first seismic waves to arrive after an earthquake are the _____.

5. The layer of soft rocks that makes up the top of the mantle is called the _____.

HBJ material copyrighted under notice appearing earlier in this work.

151

Name _____ Class _____ Date _____

CHAPTER 19 Concepts
The Changing Earth

In each space provided, write the term that best completes the statement.

1. An _____ is caused by the shifting and breaking of rocks deep underground.

2. The _____ may be near the surface or hundreds of kilometers under ground.

3. The _____ is at the surface, directly above the focus.

4. Earthquakes produce three types of seismic waves that travel along the surface of the earth and can be detected by a _____.

5. An earthquake of magnitude 3 is _____ more destructive than one of magnitude 2.

6. If an earthquake with a magnitude of 9 struck your area, you could expect _____ destruction.

7. Most of the instruments used to detect earthquakes detect change in _____ of the ground.

8. An ash cloud from an erupting volcano rolls down the mountainside because it is _____ than the air around it.

9. When the volcanic plug blows up, the reduction in pressure causes the magma to become a _____.

10. Volcanoes can be beneficial in that they add _____ to the soil.

11. When two plates run into each other, a _____ results.

12. Openings in the crust are called _____.

13. When subduction occurs, a _____ results.

14. The mantle contains _____ currents that cause the crust to move.

15. Ridges result from _____.

16. The ancient sea surrounding the supercontinent was _____.

17. Plate tectonics was first proposed by _____.

152   HBJ material copyrighted under notice appearing earlier in this work.

Name _____ Class _____ Date _____

CHAPTER **20** Vocabulary | **The Atmosphere**

A. Write the correct term in the spaces accompanying each definition. Then use the circled letters to spell the name of the subject of this chapter.

1. portion of thermosphere extending to outer space (_ _ _ _ _ Ⓞ _ _ _)
2. a device that measures air pressure (_ _ _ _ _ _ Ⓞ _ _)
3. charged particles extending to 650 km (_ _ _ _ _ _ _ Ⓞ _ _)
4. extends to an altitude of 50 km (Ⓞ _ _ _ _ _ _ _ _ _ _ _)
5. extends from 50 to 85 km above the earth (Ⓞ _ _ _ _ _ _ _ _ _)
6. air that blows toward the equator (_ Ⓞ _ _ _ _ _ _ _ _)
7. oxygen molecule consisting of three oxygen atoms (_ _ _ _ Ⓞ)
8. type of pressure from the weight of the air (Ⓞ _ _ _ _ _ _ _ _ _ _)
9. the high wind circling both polar areas (_ _ _ _ _ _ Ⓞ _ _)
10. overturning of warm and cool air (_ _ _ _ _ _ Ⓞ _ _ _)
11. 75 percent of the gases of the atmosphere (_ _ _ _ _ _ _ Ⓞ _ _ _)
12. the zone at the equator with no wind system (_ Ⓞ _ _ _ _ _ _)
13. extends from 80 to 650 km above the earth (_ _ _ _ _ _ _ Ⓞ _ _ _ _)
14. the circled letters spell (_ _ _ _ _ _ _ _ _ _ _ _ _ _)

B. Write the definition for each of the following terms on the lines below.

1. Coriolis effect _____

2. polar easterlies _____
3. prevailing westerlies _____
4. radiation balance _____
5. tropopause _____

HBJ material copyrighted under notice appearing earlier in this work.

153

Name _____ Class _____ Date _____

CHAPTER

20 | The Atmosphere
Concepts

In each space provided, write the term that best completes the statement.

1. The very light _____ in Earth's early atmosphere was able to escape the earth's gravitational pull and float off into space.

2. Carbon dioxide is used by plants to create _____.

3. Sulfur dioxide and hydrogen sulfide both combine with water in the atmosphere to form _____.

4. The planet Venus has an atmosphere similar to that of the earth before _____ evolved.

5. At sea level the atmospheric pressure is said to be _____.

6. As you go higher up in the atmosphere, the atmospheric pressure _____.

7. When you are taking off in an airplane, your ears feel as if they are going to pop because of the _____.

8. In Denver, the "mile-high city," it takes longer to cook food on a stove top because the atmospheric pressure is so _____.

9. When you cook food in a pressure cooker, the food will cook faster because of the _____ pressure.

10. When you blow up a balloon, the pressure is _____ on the sides of the balloon.

11. Winds move from a _____ air pressure to a _____ air pressure.

12. Winds are always deflected to the _____ in the Northern Hemisphere.

13. Winds in the Southern Hemisphere are deflected to the _____.

14. The high air pressure at the poles is caused by very _____ temperatures.

15. The low air pressure at the equator is caused by very _____ temperatures.

16. Air at the _____ appears to be calm, but it is rising.

154 HBJ material copyrighted under notice appearing earlier in this work.

Name _____ Class _____ Date _____

CHAPTER 21
Vocabulary | Weather and Climate

Complete the sentences below about weather by using vocabulary terms from the box. Not all terms will be used.

air mass	condensation	front	relative humidity
anticyclones	cumulus	humidity	stratus
cirrus	cyclones	hurricanes	tornadoes
climate	dew point	hydrologic	warm front
cloud	evaporation	meteorology	weather
cold front	fog	precipitation	weathering

1. _____ is the state of the atmosphere at a given time and place.

2. Moisture in the atmosphere is called _____.

3. _____ is the process by which water changes into a gas at a temperature less than the boiling point of water.

4. _____ is the amount of water vapor in a given volume of air compared to the total amount that the volume of air could hold at a given temperature.

5. _____ is the process by which water vapor changes into a liquid.

6. The _____ occurs when the relative humidity is 100 percent.

7. A _____ is a concentration of water droplets.

8. _____ is a cloud that hovers at ground level.

9. Drops of solid or liquid water falling back to the earth's surface are called _____.

10. _____ clouds are puffy and look like cotton balls.

11. _____ clouds are extended cloud layers with an even base.

12. _____ clouds consist of ice crystals and are the highest clouds in the troposphere.

13. The cycle of evaporation and precipitation is the _____ cycle.

14. An _____ is a large body of air with uniform temperature and humidity.

HBJ material copyrighted under notice appearing earlier in this work.

Name _____ Class _____ Date _____

CHAPTER 21 Concepts | Weather and Climate

A. In the space provided, indicate what will happen to the water in each of the situations below.

1. Water is in an uncovered dish at room temperature. _____

2. Water vapor is in a cooling air mass. _____

3. Water droplets are in a cloud in which the temperature is rising. _____
_____

4. Water vapor is in an air mass at -5°C. _____

5. The temperatures of both a cloud and the ground fall below 0°C. _____

6. The temperature of a cloud is above 0°C while the ground temperature is below 0°C.
_____

7. Water droplets are pushed up in a cloud and then fall through a freezing zone.
_____

B. In the space provided, write the letter for the type of cloud described.

____ 1. hot air forms puffy clouds

____ 2. made entirely of ice crystals at high altitudes

____ 3. have high winds

____ 4. cover the whole sky

____ 5. fine white strings or feathers

a. cirrus
b. cumulus and cumulonimbus
c. stratus

C. In each space provided, write the term that best completes the statement.

1. If an air mass sits over Canada for several days, the air mass will be _____ and dry.

2. If an air mass sits over the Caribbean Sea, the air mass will be _____ and wet.

3. You would expect _____ from a tropical continental air mass.

4. Since warm air is less dense than cold, warm air along a _____ front slowly climbs over the cold air and forms clouds.

Name _____ Class _____ Date _____

CHAPTER 22 Vocabulary | The Oceans

Complete the crossword puzzle below using terms studied in this chapter.

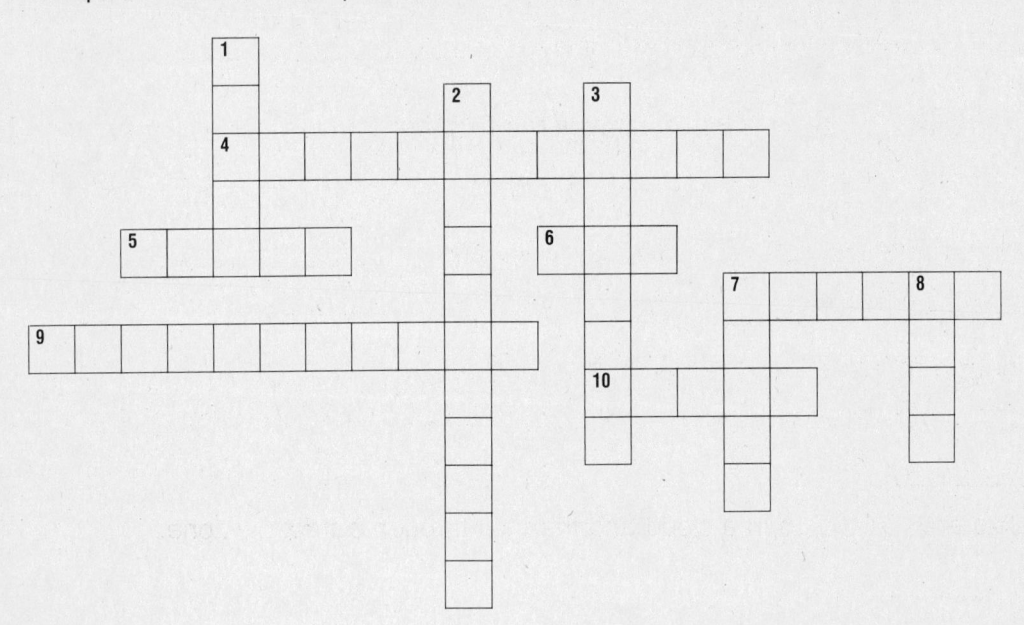

Down

1. The continental _____ marks the place where the continental shelf drops off steeply to the ocean floor.
2. High cliffs dropping to small beaches form along active _____ margins.
3. _____ is the total amount of salts dissolved in 1kg of sea water.
7. The underwater layer of land bordering the continents is called the continental _____.
8. The small tides that result when the moon and sun are at a 90° angle to each other are called _____ tides.

Across

4. The study of the ocean is called _____.
5. A _____ consists of waves several hundred meters in wavelength that travel out of storm areas.
6. When water returning to the sea forms a current flowing out along a channel perpendicular to the beach, it is called a _____ current.
7. When the moon and sun are aligned, _____ tides are produced.
9. The layer between the warm water above and the colder water below is called the _____.
10. _____ are the alternate rise and fall of the oceans and seas.

157

Name _____ Class _____ Date _____

CHAPTER 22 Concepts | The Ocean

A. In each space provided, write the term that best completes the statement.

1. A shoreline found near the boundary of two plates has _____ and small beaches.

2. Active continental margins are found where two plates collide or where _____ is occurring.

3. Beaches that are wide and sandy are found near the middle of _____.

4. For a beach to be healthy, it must have a constant supply of _____.

5. If sea level were to drop 135 m, the _____ would be exposed to the air.

6. When subduction occurs, ocean _____ are formed.

7. Shallow ocean floors are covered with _____, which is made up of clay and the shells of sea creatures.

8. The remains of _____ can be found in red clay sediments.

9. The closer you are to the midocean ridges, the _____ the layers of sediments are.

10. Where the seafloor is very old, the layers of sediment are _____ than where the seafloor is young.

B. In the spaces provided, answer the following questions.

1. If 1 kg of sea water were allowed to evaporate, what would be left? _____

2. What happens to the salinity of the ocean when there is plenty of rainfall? _____

3. In areas where there is more evaporation than precipitation, how is the salinity affected? _____

4. How did the salts from the land get into sea water? _____

5. A heat source gives off heat, and a heat sink absorbs heat. When is the ocean a heat source for the atmosphere? _____

HBJ material copyrighted under notice appearing earlier in this work.

Name _____ Class _____ Date _____

CHAPTER **23** Vocabulary | **Living Things**

A. Complete the sentences below using vocabulary terms from the box. Not all terms will be used.

```
absorption    class        nucleus
cells         cytoplasm    organelles
chlorophyll   digestion    phyla
```

1. The basic units of structure in all living things are _____.

2. _____ are structures inside of cells.

3. Between the nuclear envelope and cell membrane is a region of the cell called _____.

4. _____, a green pigment, is found in plants.

5. _____ is a process by which large, complex food particles are broken down into simple ones that an organism can use.

6. _____ is the movement of nutrient molecules through a membrane.

7. Kingdoms are divided into groups called _____.

B. Write the letter of the term that best fits each description.

_____ 1. cells are the basic units of structure and function in all living things

_____ 2. regulates the activity in a cell

_____ 3. helps support the cell but is not a living part of it

_____ 4. used to store water, waste material, or food in the cell

_____ 5. covers the body or provides a lining for its parts

_____ 6. joins, supports, or protects other tissue types

a. cell membrane
b. cell theory
c. cell wall
d. connective tissue
e. epithelial tissue
f. muscle tissue
g. nucleus
h. vacuoles

Name _____ Class _____ Date _____

CHAPTER 23 Concepts | Living Things

A. In the spaces provided, answer the following questions.

1. How would you describe the movement of a muscle cell? _____
2. How does a nerve cell respond? _____
3. How is energy released from digested food? _____
4. How do the mitochondria in any cell use oxygen to release energy? _____
5. Why do most cells reproduce from time to time? _____

B. In each space provided, write the term that best completes the statement.

1. Plant cells have cell membranes and a _____ that help support the cells.
2. Plastids act as _____ for a cell.
3. Sunlight is turned into sugar by _____.
4. Because _____ have chlorophyll, they can make their own food.
5. The animal kingdom can be broken down into two main groups, depending on the presence or absence of a _____.
6. Plants with roots, stems, and leaves are _____ plants.

C. In the spaces provided, answer the following questions.

1. What are the characteristics of a Moneran? _____
2. If a classmate showed you a single-celled organism under the microscope, how could you tell which kingdom it belonged to? _____

Name _____ Class _____ Date _____

CHAPTER 24 Vocabulary | Continuity of Life

Unscramble each term in column II, and write it in the blank to the left of its definition in column I.

I   **II**

1. _____: cell division in which the original cell splits into two cells

2. _____: rod-shaped body that carries hereditary information from one generation to the next

3. _____: results in the formation of two daughter cells that are exact copies of the original cell

4. _____: cell active only in reproduction

5. _____: process in which two gametes fuse to produce one cell that develops into a new individual

6. _____: carry traits that are passed on from one generation to the next

7. _____: strong factor in a trait

8. _____: weak factor in a trait

9. _____: genes on an individual's chromosomes

10. _____: what an individual looks like as a result of the genes on a chromosome

11. _____: type of trait in which both genes are dominant or both are recessive

12. _____: type of trait in which one gene is dominant and one is recessive

13. _____: a change in chromosome structure or in a gene

1. yrbnai nsfiosi

2. meorhcmoos

3. simsoti

4. etmage

5. lxause niderpurctoo

6. seneg

7. tnamodin

8. evcrsesie

9. epoytgne

10. epnhpoyte

11. erup

12. diryhb

13. notumita

161

Name _____ Class _____ Date _____

CHAPTER 24 Concepts | Continuity of Life

A. In each space provided, write the term that best completes the statement.

1. Simple organisms produce new cells by splitting into _____ cells.

2. Chromosomes must _____ before cell division so that each new cell receives the same genetic materials.

3. Our bodies need a constant supply of new cells because cells become _____.

4. Mitosis results in the _____ of an individual.

5. If two cells are exact copies of each other, they are the result of _____.

B. In the space provided, indicate what phase of mitosis is being described.

1. Chromosomes separate and move to opposite ends of the cell. _____

2. Chromosomes become short and thick. _____

3. Cytoplasm completely divides. _____

4. Nuclear membranes disappear while the chromosome pairs cluster at the cell's center. _____

5. The cell membrane pinches together to form two new cells. _____

C. Read about the experiment below; then answer the questions that follow.

Georgette planted 20 long-stemmed, purple flowers. She cross-pollinated them with short-stemmed, white flowers. She then gathered and planted all the seeds. All of the offspring had long-stemmed, purple flowers. Georgette planted the second generation of seeds. When these seeds developed into plants, 25 percent of the flowers were white, although there was no connection between the short stems and the white flowers.

1. If the flowers had both stamen and pistils on the same plant, how would they normally be pollinated?

_____

2. Which of the traits are dominant?

_____

3. Which of the traits are recessive? _____

Name _____ Class _____ Date _____

CHAPTER **25** | **Evolution and Natural Selection**
Vocabulary

A. Write the correct term in the spaces accompanying each definition. Then use the circled letters to spell the meaning of the name Homo erectus.

1. first humans to formally bury their dead (_ _ _ _ _ Ⓞ _ _ _ _ _ _)

2. adaptations and changes over time (Ⓞ _ _ _ _ _ _ _ _)

3. chemicals used to control insects (_ _ _ Ⓞ _ _ _ _ _ _)

4. type of selection in which those best adapted to their environment will be the most successful at surviving and reproducing (_ _ _ _ Ⓞ _ _)

5. type of breeding used to produce new breeds with the best qualities from parent breeds (_ _ _ _ Ⓞ _ _ _ _)

6. a scientist who studies fossils (_ Ⓞ _ _ _ _ _ _ _ _ _ _ _ _)

7. the most direct ancestors of today's humans (_ _ _ _ Ⓞ _ _ _ _ _ _)

8. breeding two closely related individuals (_ Ⓞ _ _ _ _ _ _ _)

9. early ancestors of modern humans (Ⓞ _ _ _ _ _ _ _)

10. Homo sapiens means wise (_ Ⓞ _ _ _)

11. the circled letters spell (_ _ _ _ _ _ _ _ _ _)

B. Write the definition for each of the following terms on the lines below.

1. beefmasters _____
_____

2. imprint fossil _____
_____

3. mildew _____

4. mold fossil _____

163

Name _____ Class _____ Date _____

CHAPTER 25 Concepts | Evolution and Natural Selection

In each space provided, write the term that best completes the statement.

1. Many of the earliest forms of life did not leave fossils because they were made of _____ parts.

2. The _____ fossils are found at the bottom of the Grand Canyon because the layers there were formed first.

3. Changes that occur over long periods of time and create new traits in an organism are _____ changes.

4. If one bird is more successful in hunting for food and in reproducing, it will have a better chance of _____.

5. If a dog breeder wanted to produce a small, short-haired, nonshedding dog and chose three different species to breed, this would be an example of _____.

6. To develop a larger and more fruitful holly tree, _____ would be used.

7. If a few insects that are resistant to pesticides are allowed to mate, the next generation will be more _____.

8. Early human ancestors are called _____.

9. A female *Australopithecus afarensis* whose fossil remains have been dated to about 3.6 million years ago is called _____.

10. Human ancestors that used tools were _____.

11. The first human ancestors that walked in a fully upright position were _____.

12. The first group to bury their dead with tools and other objects were _____.

13. The first group to create artwork were _____.

14. Modern humans are called _____.

Name _____ Class _____ Date _____

CHAPTER

26

Vocabulary

Plants and Animals

Complete the sentences below about plants and animals by using vocabulary terms from the box. Not all terms will be used.

```
amphibians      flowers          placenta
angiosperms     gymnosperms      sexual
anther          invertebrates    stamens
bony            mammary glands   stigma
cartilage       nonvascular      vascular
cones           ovary            vertebrates
coral           pistil
```

1. Plants with no internal system to move water and nutrients are _____ plants.
2. Plants with a system of interconnecting tissues used to transport water and nutrients are _____ plants.
3. Evergreen trees are the best-known kind of _____.
4. Evergreen trees produce their seeds in _____.
5. Roses, lilies, and carnations are types of _____.
6. Roses, lilies, and carnations produce their seeds in _____.
7. The male flower parts are called _____.
8. The female part of the flower is the _____.
9. The part of the flower that produces the pollen is the _____.
10. The pistil has a sticky part called the _____.
11. The base of the pistil contains the _____ with the eggs.
12. Animals that have a backbone are called _____.
13. Animals with no backbone are called _____.
14. Nearly all the world's fish belong to a class called _____ fish.
15. Vertebrates that spend part of their lives in water and part on land are _____.

HBJ material copyrighted under notice appearing earlier in this work.

Name _____ Class _____ Date _____

CHAPTER 26 Concepts | Plants and Animals

A. In the space provided, indicate the type of plant described.

1. a tree with needle-shaped leaves, pollen cones, and seed cones _____

2. a broad-leaf tree that produces a dry, acorn type of fruit _____

3. a short, soft plant that grows in damp areas and reproduces by alternations of generations _____

4. a broad-leaf tree with brightly colored flowers that develop into fruits _____

5. a tiny, heart-shaped structure on which the gametes develop _____

B. In the space provided, indicate which group or groups of vertebrates have the following traits.

1. placenta _____

2. eggshells to protect young _____

3. warm blooded _____

4. skin that absorbs oxygen _____

5. only two legs _____

6. must reproduce in water _____

7. young are fed milk _____

8. four-chambered heart _____

9. covering of feathers _____

10. covering of hair _____

11. eggs are fertilized inside female _____

12. scales _____

Name _____ Class _____ Date _____

CHAPTER 27
Vocabulary | Human Biology

Complete the crossword puzzle below using terms studied in this chapter.

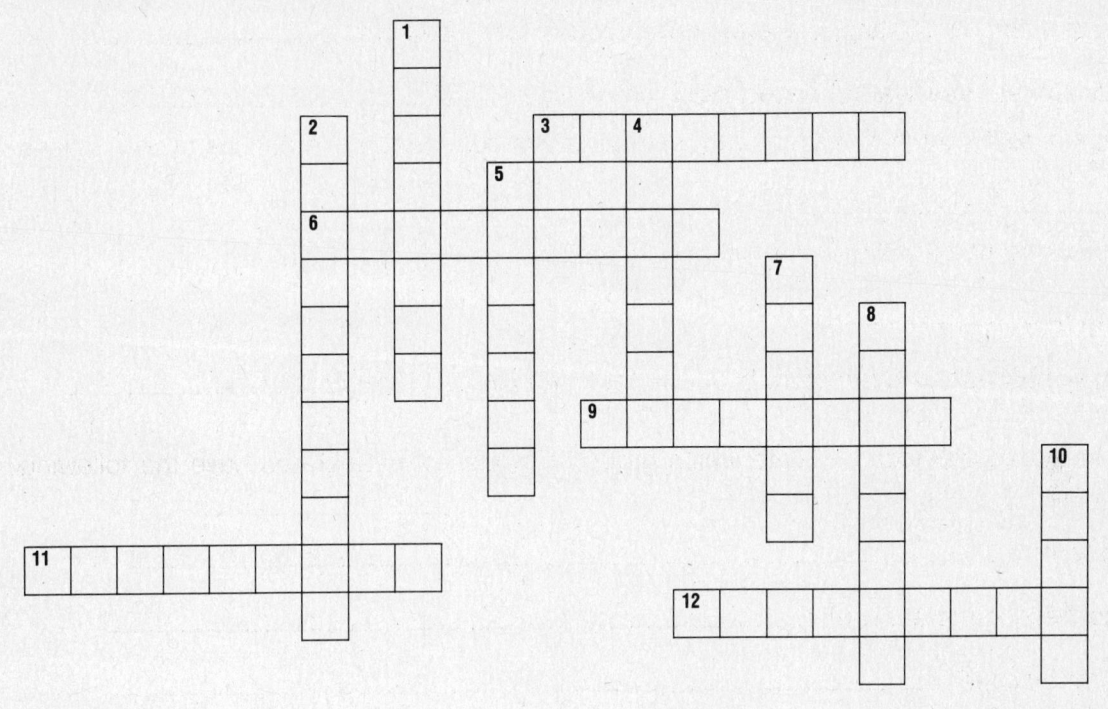

Down

1. a stimulant found in coffee, tea, chocolate, and many soft drinks
2. the smallest blood vessels
4. an illegal stimulant usually found in powder form
5. chemicals that help in the digestion of food
7. the cord that is the link between the brain and all other parts of the body
8. the system that works with the skeletal system to move bones
10. vessels that carry blood back to the heart

Across

3. found in tobacco, increases blood pressure, and causes the heart to beat faster
6. the condition that results when an egg and sperm unite
9. the highest level of health a person can achieve
11. a disease that prevents the lungs from exchanging carbon dioxide and oxygen
12. the release of an egg from a female's ovary

Name _____ Class _____ Date _____

CHAPTER 27 Concepts | Human Biology

A. In each space provided, write the term that best completes the statement.

1. The main functions of the skeletal system are to provide _____ for the body, to _____ the soft body tissues, and to make _____.

2. The muscular system's function is to move the _____.

3. The _____ system's function is to coordinate the movement of the bones and muscles.

4. The heart, blood vessels, and blood all make up the _____.

5. The _____ make antibodies that destroy harmful organisms.

6. The function of the respiratory system is to take _____ to the lungs and remove _____ from the body.

7. The job of the _____ system is to remove liquid waste from the body in the form of _____.

8. The job of the reproductive system is to allow two individuals to make a new _____.

9. The growth and development of an individual depend on secretions of the _____.

10. The digestive system _____ breaks down food by use of teeth and muscles.

B. In the space provided, indicate the human body system that includes each of the following.

1. stomach _____

2. brain _____

3. ligaments _____

4. marrow _____

5. kidneys _____

168

Name _____ Class _____ Date _____

CHAPTER
28
Vocabulary | **Ecology**

A. Complete the sentences below using vocabulary terms from the box. Not all terms will be used.

abiotic	consumer	ecosystem
biome	decomposers	estuary
biotic	desert	niche
community	ecology	population

1. A _____ consists of all the living organisms in an ecosystem.

2. The nonliving factors that affect an ecosystem are called _____ factors.

3. The living, or _____, components of ecosystems include all living things.

4. A _____ is a large area of similar climate and vegetation.

5. An organism that does not make its own food is called a _____.

6. The study of how living and nonliving things interact is _____.

7. Organisms that release nutrients from dead material are called _____.

8. An _____ is a group of living and nonliving things interacting over time.

9. The _____ receives less than 20 cm of moisture per year.

10. An _____ is a shallow area where fresh water and salt water mix.

B. Write the letter of the term that best fits each description.

_____ 1. a group of organisms of a single species, living in a certain area

_____ 2. a population's ability to grow in an environment and reproduce itself without limits

_____ 3. the sum of all the limiting factors

_____ 4. movement of new individuals into an area

_____ 5. change in ecological communities over time

a. biotic potential
b. environmental resistance
c. immigration
d. population
e. predators
f. scavengers
g. succession

169

Name _____ Class _____ Date _____

CHAPTER **28** Concepts | **Ecology**

In each space provided, write the term that best completes each statement.

1. Ecosystems are made up of living and _____ things.

2. Climate, sunlight, and soil are _____ factors that affect the number and kinds of organisms that live in an area.

3. All the squirrels in Central Park make up a _____.

4. One abiotic factor that would affect the size of a group is _____.

5. A biotic factor that affects the squirrels in Central Park is _____

_____.

6. The _____ that the squirrels live in includes birds, cats, chipmunks, dogs, people, grass, trees and bushes.

7. The squirrels live in the park; the park is their _____.

8. The _____ of the squirrels is to eat acorns to limit the growth of oak trees.

9. If a new group of squirrels is released into the park, the population increases due to

_____.

10. If a large number of squirrels die because of a bacterial infection, the population

_____.

11. If the squirrels in the park eat all the acorns available, the lack of food will be a

_____ for the population size.

12. Limiting factors that keep the earth from becoming populated with one type of organism are

_____.

13. Once the squirrel population reaches its carrying capacity, further population increases will

damage the _____.

14. Energy on Earth comes from the _____.

15. Chemicals must be _____ because there is no new source for them.

16. Nitrogen gas is changed into nitrates by _____.

170 HBJ material copyrighted under notice appearing earlier in this work.

Name _____ Class _____ Date _____

CHAPTER **29** Vocabulary | **Managing Natural Resources**

A. Unscramble each term in column II, and write it in the blank to the left of its definition in column I.

I

1. _____: long period without precipitation
2. _____: loosening and movement of soil from one place to another by wind and water
3. _____: all the water in oceans, rivers, lakes, streams, and ponds
4. _____: water found beneath the earth's surface
5. _____: convert solar energy to electricity
6. _____: species no longer in existence
7. _____: number of individuals in a species is low and in danger of going out of existence

II

1. hurdogt
2. nosoire
3. ecasfru reawt
4. dnguro trawe
5. iteethpoolcrc eclsl
7. tctinxe
8. dredenneag

B. Complete the sentences below by using vocabulary terms from the box. Not all terms will be used.

multiple use management	nuclear fusion	solar energy
nonrenewable resources	renewable resources	sustained yield management
nuclear fission		

1. Air, water, plants, and animals are examples of _____.
2. Minerals and petroleum are examples of _____.
3. Harvesting lumber at a rate equal to the rate at which timber is being replaced is called _____.
4. Using land for many different purposes is _____.
5. Energy received directly from the sun is called _____.

HBJ material copyrighted under notice appearing earlier in this work.

171

Name _____ Class _____ D___ _____

CHAPTER **29** Concepts | **Managing Natural Resources**

A. Complete the crossword puzzle below using concepts studied in this chapter.

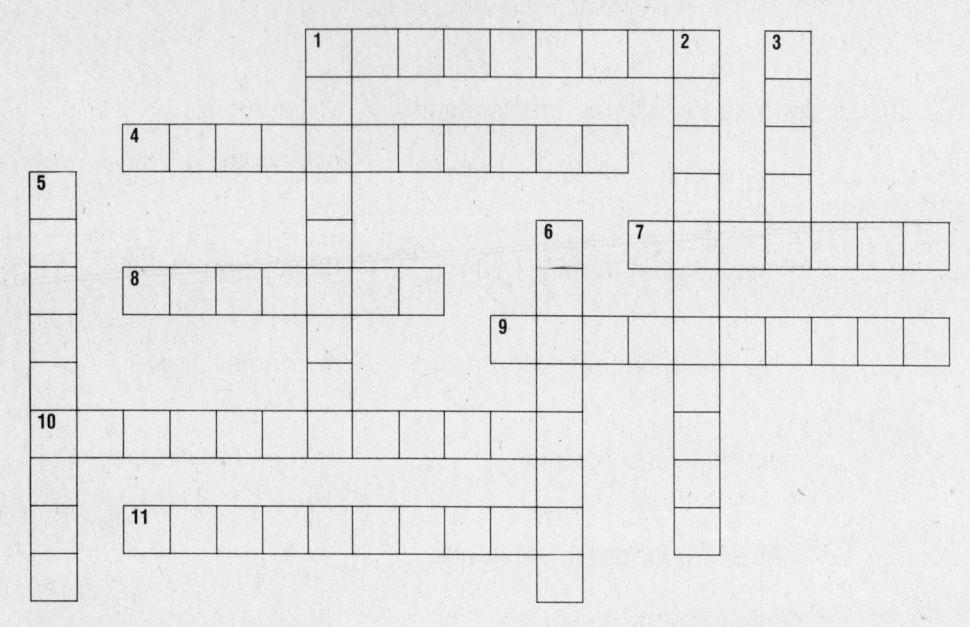

Down

1. resources such as air, water, plants, and animals
2. must be concentrated to be useful to humans
3. determines what types of plants will grow
5. slowly release ground and surface water
6. inorganic substances that cannot be renewed

Across

1. air, water, land, animals, plants, and minerals
4. valuable in dry climates
7. results from the action of wind and water
8. living matter and its products
9. blue whale, grizzly bear, leopard
10. oceans, rivers, lakes, streams, ponds
11. steam from the earth

B. In the space provided, answer the following questions.

1. Explain the difference between fission energy and fusion energy.

_____

_____

2. What are some ways in which fossil fuels can be conserved?

_____

_____

172